Study Guide for Lamanna and Riedmann's

Marriages and Families
MAKING CHOICES AND FACING CHANGE

FIFTH EDITION

DAVID TREYBIG
Baldwin-Wallace College

Wadsworth Publishing Company
Belmont, California
A Division of Wadsworth, Inc.

International Thomson Publishing
The trademark ITP is used under license.

Printer: Malloy Lithographing, Inc.

Printed in the United States of America

1 2 3 4 5 6 7 8 9 10—98 97 96 95 94

ISBN 0-534-18740-4

Contents

PREFACE
How to Use the **Study Guide**

This **Study Guide** contains two sets of instructions about how to use the **Study Guide.** For people who know how to study and how to use a study guide, one set of instructions is very brief. For people who want more detailed information about how to study and how to use the **Study Guide,** the other set of instructions is more detailed. Feel free to read both sets of instructions if you wish. Most people find the instructions to be quite helpful. Some report that their grades improve one or two letter grades when they follow the suggestions below.

BRIEF INSTRUCTIONS

Although there is no foolproof way to study, to learn, and to prepare efficiently to take examinations, the following suggestions may assist you.

1. <u>Before</u> you read the chapter, examine the **Chapter Overview** and **Chapter Summary** in this **Study Guide.** This will give you some idea of what the chapter is about. Armed with this intellectual road map, you will find the chapter easier to understand and remember.

2. Read the chapter section by section, <u>pausing</u> <u>between</u> <u>sections</u> in order to <u>think</u> about what you have read. Please do not read the entire chapter without pausing, because you might have difficulty retaining what you have read. <u>Read</u> a section, and then <u>stop</u> <u>to</u> <u>think</u> about what you have read. When you feel comfortable that you know and understand the material, repeat the process until you have completed reading the chapter. Studying like this may take more time, but it tends to produce more learning and better grades.

3. Be sure to pay attention to the charts, graphs, case studies, and even the written material accompanying the illustrations and photographs. The instructor may include questions on this material. If the material was not important, it would not be included in the text. But don't memorize it. Understand it.

4. After reading the chapter as indicated above (some students like to make notes as they read), reread the **Chapter Overview** and **Chapter Summary** in the **Study Guide**. Then answer the questions in the **Study Guide**. After you have answered them, check the answer using the answer key at the end of each **Study Guide** chapter. If you miss a question, review the material in the text to find out why you missed it.

5. Remember, in a course like this one, one of the goals is to <u>understand</u> the material. Don't memorize it. <u>Understand it.</u>

6. Finally, pause between studying this material and going on to another task. And remember to allow for a thorough review before taking examinations.

DETAILED INSTRUCTIONS

This **Study Guide** is intended to help you study <u>**MARRIAGES AND FAMILIES**</u>, 5th Edition, by Lamanna and Riedmann. The **Study Guide** can give you helpful suggestions. It can give you a brief overview of the chapter and alert you to important terms, concepts, theories, and research results. It can even give you some sample exam questions so that you can test your understanding of the text. But the **Study Guide** is not a substitute for the text. Instead, it complements the text and properly used should help you to understand the text better and also help you make better grades

on exams. Ultimately, each student must find her
or his own ways of effective study. But the
suggestions below have proved very effective for
many students.

FORMAT OF THE STUDY GUIDE

The study units in the **Study Guide** follow the
format and organization of the text, <u>**MARRIAGES AND**
FAMILIES,</u> 5th Edition, by Lamanna and Riedmann.
Just as in the text, the **Study Guide** contains a
study unit for the **Prologue,** for each of the
sixteen chapters in the text, and for the
appendices.
Each study unit contains the following:

1. A **Chapter Overview** that briefly notes the
    topics to be covered in the chapter.

2. A **Chapter Summary** that captures the
    essential points in a few paragraphs.
    Because of the superior quality of the
    text's chapter summaries, the **Chapter**
    Summaries are based primarily on the
    text's summaries.

3. **Key Terms** that should be fully
    understood. The student should fully
    understand each key term and be able to
    define it using his or her own words as
    well as using technical sociological
    terms. The student should be able to
    give one or more examples of each key
    term and explain <u>why</u> the example is an
    example of the key term.
        Of course, the text routinely uses
    bold type to indicate new terms when the
    terms are first used, and these should be
    understood. But the **Key Terms** in the
    Study Guide are central, key, important
    terms in the chapter and as such should
    be given special attention.

4. **Completion Items,** or sentences in which
    the student writes in the appropriate key

term from the list of key terms given previously in each chapter.

5. **Key Research Studies.** It is critical that students understand the findings of important research. The **Key Research Studies** alert students to the research to which they should pay especially close attention in the text. Though students should be alert to all research findings discussed in the text, the **Key Research Studies** should be closely examined regarding the purpose of the research, how the research was conducted, what was found, and what the authors think the research means in terms of marriages and families.

Typically, the names of the researchers are given along with their research. Always, however, the most important thing to be learned is what the researchers found, not the names of the researchers. If you can recall the names of the researchers, that is admirable and of course desirable. But the essential thing to be learned is what the researchers discovered. The sample test questions will help you to develop this study skill.

6. **Key Theories.** The main purpose of theories is to help us to understand why things are as they are. Though there may be other theories in the text chapters, the **Key Theories** are central to understanding the chapter material. Therefore, students should be certain that they have a complete understanding of the **Key Theories.**

Here, too, the most important things to learn are the name and the meaning of the **Key Theories,** not the name of the persons who proposed or developed the theories. Again, it is admirable and desirable if you can remember the theorists' names, but the essential thing is to remember what the theory is called and understand what it means.

7. **True-False Questions.** These questions are included in the **Study Guide** to help you test your knowledge of the text material and because some instructors include such questions on examinations. The true-false questions in the **Study Guide** are not intended to be "tricky." They are written to be obviously true or obviously false.

8. **Multiple-Choice Questions.** These questions are included for the same reasons the true-false questions are included, to allow you to test your knowledge of the chapter material and because some instructors use multiple-choice questions on examinations.

9. **Short-Answer Essay Questions** and **Essay Questions** are included in this study guide. The student should be able to answer both types of essay questions because instructors may use these questions or questions like them when constructing an examination.

    In addition, answering these questions can be an excellent way to review the chapter material. Many students write out the answers to these questions and then review their written answers before taking examinations. It can be an effective technique.

10. **Answers to completion items, true-false questions, and multiple-choice questions.** If you miss questions, you may not fully understand the chapter materials. Likewise, if you are correct in answering these questions, you probably have a knowledge and understanding of the chapter material. Although you will probably not get the **Study Guide** test questions on your examinations, many instructors use questions similar to the questions in the **Study Guide**. Therefore, do not study the **Study Guide** questions; instead, study the text. The **Study Guide** can help you, but don't assume that these questions are the only questions that

will be on your actual exams or tests. **Exams will have other questions as well!** Study the text, and then use the **Study Guide** questions as a <u>rough indicator</u> -- at least at the time you're using the study guide--of how effective your studying has been.

SUGGESTIONS FOR EFFECTIVE STUDY SKILLS:
Studying the Test Smarter, Not Harder

How sure are you that your study skills are good ones? Are you sure that they are the best they can possibly be? Can they be improved? It is now recognized that the great majority of students have poor study skills. Most students who study but still get low grades get them because their study skills are low. If you spend considerable time studying but don't do as well on examinations as you would like, then your study skills can probably be improved.

You might consider the following suggestions as ways to improve the effectiveness of your study time.

Study Time

How much time should one spend studying? It differs from person to person and from course to course, but a useful rule of thumb is that one should spend from one to three hours studying for every hour spent in class. Most students seriously underestimate the amount of time that needs to be spent studying in order to do well in a course.

The text does not need to be merely <u>read.</u> It needs to be <u>studied.</u> And studying it is very different from just reading a chapter from beginning to end without pause. To study the text adequately takes time. Likewise, taking class notes does little good if they are never studied. And merely reading through class notes is not equivalent to studying them.

Realistically appraise your personal schedule and set aside sufficient study time for each course you are in. Decide on the days and the times you will study, write those study sessions

on your daily calendar, and then stick with it.
Self-discipline and assertiveness are needed in
order to keep to your study schedule, for a
schedule is necessary if you are to get the kinds
of grades and levels of learning that you want.
If friends or kin knock on your door while you are
studying and invite you to do something else, just
tell them "No," and suggest a later time when you
would be available. If your studies are
important, establish a definite schedule and stick
with it.

Studying a Text: Some Strategies That Work

Many students who have learned and used the study
skills that follow have ordinarily been able to
increase their exam grades by one or two letter
grades. If you want your grades to improve, you
can probably improve them. It is primarily a
matter of knowing how to study and then putting
that knowledge into action. Merely reading
through a text chapter from beginning to end is
not an effective way to learn, understand, and
retain the material and get better grades.   There
are specific strategies for effectively studying a
text, and if you so desire, you can learn what
these skills are and apply them.
     These skills have appeared in various
combinations and have been given various names
(SQ3R or SQ4R, for example). But regardless of
what they are called, the basic skills are the
same. If you learn them and use them, you will
probably see a significant increase in your level
of learning and in your grades. It is not so much
a matter of having high intelligence as it is a
matter of applying easy-to-learn methods of
studying the text.

Study Guide Exercises and Exam Questions

If you can answer the **Study Guide** questions as
well as the **Study Questions** at the end of each
chapter, you ought to do very well on

examinations.   There are several reasons for this:

1. The **Study Guide** questions probably will not be identical to the questions you get on an examination, but they will be similar to many of the examination questions.

2. Think of the **Study Guide** questions as a sample of the questions you may have on examinations. If you do well on the **Study Guide** questions, then you should probably do equally well on the examination. On the other hand, if you do not do very well on the **Study Guide** questions, then you may not do very well on the examinations unless you raise your level of learning and understanding.

3. The **Study Guide** can alert you to specific topics about which you should be familiar. If you miss a question, you can "repair the damage" by reviewing the material in the text.

Testing Your Knowledge

The **Study Guide** can be used in several ways to test your knowledge of the material in the text. If you use all of them, the results should be, first, insight about how well you know the text material and, second, additional learning as you go through the **Study Guide** process.

Key Terms

Begin testing your knowledge with the key terms listed for each chapter in the **Study Guide**. Turn each key term into a question--and then try to answer it. If the key term is "life spiral model," ask yourself "What is the life spiral model?" and try to answer the question. If you cannot answer the question or if you feel unsure of your answer, review the material in the chapter. Continue in this manner until you have turned all of the key terms into questions and, it is hoped, into correct answers. (Incidentally, some students find it very helpful to explain the answer to such questions to another person, preferably to another student in the class. If you just think the answer through, you can sometimes fool yourself into thinking that you know the answer when you don't. But if you have to explain the answer to another person, it becomes clear whether or not you know the material. You can use this technique on other types of questions as well.)

Completion

Next, answer the completion items. Read each sentence and fill in the blank with the correct term from the **Key Terms** section. Read each sentence, scan the list of **Key Terms**, and write what you take to be the correct answer in the blank space. When you have done this for all of the completion items, check your answers by referring to the answers at the end of each **Study Guide** chapter.

If you missed one or more of the completion items, review the material in the text chapter. Later, when you are reviewing for an examination, it can help to review your completed answers in this section of the **Study Guide.**

Key Research Studies and Key Theories

For the **Key Research Studies** and **Key Theories** sections, turn each research study and each theory into a question that you ask yourself and then try to answer. If the **Key Research Studies** section lists Geerson's research on women's role choices, ask yourself what it was that Geerson found when this research was conducted. You should know.

If the **Key Theories** section lists family development theory, ask yourself what family development theory is, what questions it addresses, whether it has any particular strengths and weaknesses, and so on. Again, whether you tell the answers to such questions to another person (as is recommended), or just think the answer through, or write the correct answer, be sure to review the material if you don't know the answer or if you feel insecure about the answer.

True-False

The true-false questions are written so that they are obviously true or obviously false. None is intentionally tricky. Take each question just as it is. Do not add or subtract anything from the question in your own mind. Is the statement true, according to the text? Or is it false?

Print T or F in the blank space to the left of each true-false item. When you have completed all of the items, check your answers with the true-false answers section at the end of each chapter in the **Study Guide**.

If you miss one or more true-false items, find the material in the text to discover why the answer is other than you thought. If you miss a question, be sure to print the correct answer in the **Study Guide** so that you can use these questions as part of your review process. (When reviewing, cover the correct answers with a sheet of paper or card, and you can speedily check your responses.)

Multiple Choice

When you answer a multiple-choice item, do not read the entire question without pause. Instead, read the first part of the item (the "question" part) to locate what the item is about. Before continuing through the alternative answers, stop for a moment and think about that topic. You might want to briefly think about what the correct alternative answer should be. Then read the alternative answers. If the item is well constructed and if you know the text material, your task is easy: Select the correct answer.

Well-written multiple-choice items can be an excellent test of your knowledge of the material. In a well-constructed item, the correct alternative answer should not be obvious or stand out because it is "different" from the others. The student should not be able to infer the answer using only logic. To a student who does not know the text material, all of the alternatives should seem reasonable. To the student who does know the text material, the correct alternative answer should be apparent and the other alternatives should be plausible but incorrect according to the text.

Print the correct alternative answers in the spaces to the left of the multiple-choice questions. After you have answered all the questions, check your answers with those in the answer section.

Your level of correct answers for the **Study Guide** questions is a fairly accurate estimate of your level of knowledge of the text material. If you missed a question, return to the chapter and reread the material so that when you review the multiple-choice questions you will answer them (and questions similar to them on the examination) correctly.

Essay

The **Study Questions** at the end of each chapter in the text are excellent to use to practice writing answers to essay questions. If you understand the chapter, you should be able to answer these questions or others like them. Answering them will help you to see the overall issues and questions being addressed by the chapter.

CONCLUSION

A study guide cannot work miracles, especially if it is not used properly. But this **Study Guide**, properly used, should help the student to learn more and do better on examinations. If the **Study Guide** is used correctly and its suggestions are followed, the student should expect to increase his or her knowledge of **MARRIAGES AND FAMILIES** and do well on examinations covering text material.

Good luck with our academic efforts, and remember:

Give yourself enough time to study.

Study because you **want** to, not because you **have** to.

Be interested in the material or find a way to **become** interested in it. One way to become interested in the material is to try to apply it to yourself and to your daily life.

Don't **memorize** the material. **Understand** it!

Acknowledgments

I would like to thank the authors for asking me to write the **Study Guide** for **MARRIAGES AND FAMILIES,** 5th edition, by Lamanna and Riedmann. I also want to thank Jennie Redwitz, Susan Shook, and Serina Beauparlant, for editorial production assistance and support.

At Baldwin-Wallace College, Jane Brown and Helen Rizzo assisted by proofreading and by providing constructive suggestions from a student perspective. Carol Corpus assisted in many ways by coodinating activities and by making available needed equipment. Baldwin-Wallace College provided a variety of necessary support for this project. Christopher Sullivan, Director of Computer Services, continues to be an invaluable source of ideas, encouragement, and support.

David Treybig, Ph.D.
Department of Sociology
Baldwin-Wallace College
Berea, Ohio   44017

PROLOGUE
**Marriages and Families: Making Choices
and Facing Change**

Prologue

The Prologue discusses some choices adults make
over the course of their lives. Although no two
people's life trajectories are exactly the same,
Americans tend to share certain periods of
stability and transition. The most important idea
in the Prologue is that adults change. Because of
this, marriages and families are not static.
Every time one individual in a marital
relationship changes, the relationship changes,
however subtly. The text examines some creative
ways in which mates can alter their relationship
in order to meet their changing needs.

Chapter Summary

People change. We all know that children change.
But adults change as well. People make choices as
they are influenced by a number of factors,
including _____ expectations, race and
ethnicity, religion, social class, gender, and
_____ events. In these ways people are often
different. But in many other ways they are
_____, sharing some common attitudes about
work, education, marriages, and families.
    Individuals and families are similar in many
ways to other individuals and families. But they
are also in many ways unique and like no others.
It is for this reason that social scientists
measure and discuss factors of similarity--such as
the typical family life _____, in which the main
theme is living life in terms of the typical or
usual sequence expected in our society--and
dissimilarity or uniqueness embodied in the highly
varied ways people make various choices in their
lifestyle. Sometimes people choose knowledgeably.
That is, they are actively involved in assessing
their options, their preferences, and other
aspects of their life situation--and choose.
Others may not be actively involved in making
decisions or choices and simply let things happen

to them, enduring with varying levels of success whatever comes their way. This is also a way of choosing--<u>choosing by</u> _____.

No two people are exactly alike, yet they are alike in some ways. Adults change as they move through the family life cycle, yet there are factors that persist even as this change occurs.

The variety of life-style choices adults can make today is what is meant by the <u>life</u> <u>spiral</u> model of adult change. Some adults follow fairly traditional patterns, some incorporate alternative patterns into their otherwise traditional life, some shift from one pattern to another, and so on. But sometimes two individual adults move in their own ways, at their own paces, making choices and decisions--some traditional, some nontraditional-- in ways that may affect one another. Change,

then, is one of the main perspectives guiding the ideas set forth in this text.

> **Point to Ponder:** Which of the following best describes your life so far: life cycle or life spiral? Explore thoughtfully why you answer as you do.

Key Terms

You should be able to explain the concepts listed below. In your explanation, try to avoid using the concept you are explaining. You should be able to give several examples of each concept and to explain <u>why</u> each example is an example.

age expectations
  (p. 6)
choosing by default
  (p. 7)
choosing knowledgeably
  (p. 8)
life spiral
  (p. 9)
family values (familism)
  (p. 10)

archival family function
  (p. 10)
individualistic (self-
  fulfillment) values
  (p. 10)
family life cycle and
  its stages
  (p. 13)

Complete the following sentences by selecting the correct alternative from the terms listed above. Some may be used more than once. Some may not be used at all. Filling in a blank may require more than one word.

1. Andrea wants to become a college freshman now that her children have graduated from college. Her friends tell her she ought to prepare for grandchildren, not for college classes. This illustrates _____.

2. A student secretly worries about whether or not going to college is the right thing to do at this point in life. While pondering, time goes by, classes and exams are missed, grades decline, and the student is suspended from school. This is an example of what the text calls _____.

3. Tom pondered for two weeks whether or not to work part-time while being a full-time student. He considered the positive and negative aspects of his options, and then made a decision, fully prepared to live with the consequences. This illustrates what the text calls _____.

4. Sara Lee often surprises her friends by her choices, many of which are nontraditional or innovative. Her life, conventional in some ways, takes interesting and unexpected turns, the results of choices she makes. This illustrates the concept of _____.

5. When couples value spending time together in shared activities, perhaps value having or adopting children, value turning to this intimate group as a source of many satisfactions, then family values or _____ are expressed.

6. Having yearly family reunions at which attendance is stressed, acquaintances are renewed, happy memories are relived, and photographs are shared is an example of the _____.

7. Frank considers marrying his long-time friend, but he decides not to because he wants to have a lot of time to himself, enjoy the things **he** wants to enjoy, likes his privacy, and wants to feel free to pursue what tempts him at the moment, with only himself to please. Clearly, Frank is a strong supporter of _____.

8. When each member of a couple tries to go her/his own way, doing innovative and creative things that are not exactly what society expects, and at the same time sustain commitment and time for their relationship, this is an example of problems and complexities in what the text calls the _____.

9. Social and cultural expectations about the various "phases" or periods of family life, along with appropriate times to enter and then move on to what life usually holds next for family members, is what the text means when it uses the phrase _____.

Key Research Studies

You should be familiar with the main question being investigated and the research results for the following studies:

    Aldous: the seven-stage family life cycle
    Etzkowitz and Stein: the life spiral model
    Weigert and Hastings: archival family functions

Key Theories

Family development theory
Life spiral model

True-False

____ 1. Social factors influence people's choices in two ways--social norms and available options.

____ 2. As the age structure of our society alters, heavy burdens of care for older people fall on middle-aged sons and daughters, but mostly on middle-aged sons.

____ 3. The fertility rate of blacks is the highest of the United States' racial/ethnic groups.

____ 4. Divorce rates are highest in the West and South, and lowest in the Northeast.

____ 5. Midwestern states' divorce rates are in the moderate range.

____ 6. People choose by default when they make a specific choice solely because it seems like the easiest thing to do.

____ 7. The life spiral model has some advantages over other models, but it fails to take into account alternative role choices in the life course.

____ 8. Familism exists in American society, but it does not permeate our culture.

____ 9. The archival family function refers to the process whereby family identity is maintained and preserved.

____ 10. In all intimate relationships--including families--it is usual for partners to alternate between separateness and togetherness.

____ 11. The life spiral concept focuses on the fact that those people who move into and out of family life cycle stages "on time."

____ 12. The text argues that an important way for individuals to keep their balance in shifting, stressful personal relationships is to clearly establish who is to blame for previous mistakes.

____ 13. In the seven stages of Aldous's family life cycle, "families with school children" is the third stage.

____ 14. In the "parents of young adults" stage of Aldous's family life cycle, the young adult children may leave and return home several times.

____ 15. According to the text, personal decision making is influenced by society, but society is not influenced by individuals' personal decision making.

Multiple Choice

1. Which of the following statements is FALSE?
   a. Social factors limit people's options.
   b. Society encourages people to select some options over others.
   c. Religion and social class pressure people to select some options but not others.
   d. Current events pressure people to make certain choices rather than others, but gender does not exercise pressure in this way.

2.   In the _____, Americans found marriage much more affordable than they do today.
     a.   1940s
     b.   1950s
     c.   1960s
     d.   1970s

3.   As recently as 100 years ago, _____ of our population died before reaching adulthood. Today three-fourths of the United States' population live to be 65, and many reach 85.
     a.   one-third
     b.   one-half
     c.   one-sixth
     d.   one-eighth

4.   Which of the following does NOT affect people's options and decisions?
     a.   race
     b.   gender
     c.   age expectations
     d.   none: all affect options and decisions

5.   One reason for the increase in female-headed black households is:
     a.   black women's preference for households with unquestioned female dominance
     b.   a decline in the number of black husbands who can help support a wife and children
     c.   black males' preference for unattached relationships
     d.   black females increasingly negative attitude toward birth control and abortion

6.   Which of these, compared to the others, has been shown to have a different balance of power in marriage?
     a.   whites
     b.   blacks
     c.   Chicanos
     d.   none of the above

7. Divorce rates are lowest in which of the following areas of the country?
   a. the West
   b. the Northeast
   c. the Northwest
   d. the South

8. Age expectations refer to which of these?
   a. experiencing specific life events at socially appropriate times
   b. expectations people usually have for their offspring at different ages
   c. the age to which people expect to live
   d. income expectations people have when they are at various ages

9. In the context of this text, choosing knowledgeably refers to making choices based on:
   a. knowledge about available alternatives and options
   b. the most factual or scientific information available
   c. those things one has learned from reliable sources of knowledge
   d. none of the above

10. Choosing by default can be summarized as:
    a. letting your personal flaws make your decisions for you
    b. single-mindedly pursuing a course of action one has decided on
    c. avoiding conscious decisions and just letting things happen to oneself
    d. selecting a course of action but remaining open to carefully considered decisions at a later date

11. An important part of choosing knowledgeably is:
    a. deciding on the basis of facts rather than values
    b. awareness of consequences
    c. not being limited in choices by any values one might have
    d. choosing from the options created by one's own generation rather than by generations past

12. Values clarification is:
    a. a laboratory-based procedure whereby the values of a group are discovered
    b. the process by which an individual becomes aware of his/her most important values
    c. another term for the socialization process
    d. the process by which societies develop the values of their own cultural system

13. Which of these is a component of decision making, according to the text?
    a. feedback
    b. didactics
    c. chipping
    d. rechecking

14. Value cleavage refers to:
    a. adopting values that conflict with those of one's closest kin
    b. the conflict between values stressing conformity to social norms and values stressing individualism
    c. one's growing awareness that making personal progress requires learning new values and skills
    d. those values that are the newest in a society, that are on the "cutting edge" of social change

15. The text states that the _____ is meant to better accommodate the variety of life-style choices people make today and to acknowledge that people do not necessarily enter into certain stages "on time."
    a. interrupted sequences theory
    b. roller-coaster theory
    c. reality-testing model
    d. life spiral

16. Life in American families reflects the tension between:
    a. individual freedom and self-orientation
    b. family solidarity and familism
    c. the family group and other groups
    d. family solidarity and individual freedom

17. Among the other things they do, families also:
    a. perform a dislocation function in society
    b. perform latent functions, which are probably families' main functions
    c. create physical and psychological boundaries
    d. none of these

18. In the Jones family, an elderly aunt is recognized by everyone in the family as the person to give things to: old photographs, letters from deceased family members, and other memorabilia. And people come to her for information about past family events that are important memories. This elderly aunt's activities are related to which of these?
    a. positive social lag
    b. life spiral record
    c. archival family function
    d. delayed socialization

19. Family life in any form:
    a. displays people's capacity to sustain interaction without changes in levels of commitment
    b. is found in all societies except the most simple, preindustrial systems
    c. provides a higher level of gratifications than does singlehood
    d. has both costs and benefits

20. An integral, central part of family development theory is the concept of:
    a. a developed, fixed personality
    b. the flow of influence between generations as a one-way flow, from grandparent to parent to child
    c. change
    d. individualism as preferable to familism

21. A challenge for contemporary relationships is to integrate divergent personal change into the relationship. Partners are able to change in important ways and integrate those changes in a way that helps sustain the relationship. This is the concept of:
    a. behavioral symbiosis
    b. coordinating partners' life spirals
    c. symbiotic interpersonal relationships
    d. dual career partners

22. Linear movement through a set of stages over time is:
    a. temporal family development
    b. phase-active family development
    c. the family life cycle
    d. adaptive life stages

23. Approximately _____ of African-Americans become parents before marrying.
    a. one-fourth
    b. one-half
    c. three-quarters
    d. four-fifths

24. Comparing African-Americans and whites, the text notes that:
    a. after divorce, African-Americans remarry more quickly than do whites
    b. whites have a significantly higher divorce rate than do African-Americans
    c. whites are likelier to live together before marriage than are African-Americans
    d. whites arrive at the postparental stage much earlier than do African-Americans

25. Which of these is **not** one of the four themes developed by the Lamanna and Reidmann text?
    a. personal decisions must be made throughout the life course
    b. we live in a changing society, and this can make personal decision making more difficult and more important
    c. personal decision making feeds back into society and changes it
    d. none of these

> **Your Opinion, Please:** Some people feel that
> it is unfair to a
> partner to enter a relationship with one set
> of understandings and agreements and then
> later--in five or ten years--to change one's
> mind and want to go in a new direction.
> What is to be done when a person wants to
> change but is in a relationship based on
> previous agreements and assumptions?

Short-Answer Essay Questions

The following are sample short-answer essay
questions--questions of the type you may be asked
if your instructor uses questions like these.
Even if your instructor does not use questions
like these, you can help organize and consolidate
your learning if you can answer these questions in
a _well-organized_ and _complete_ manner. Do not
think that brief essays are easier than lengthy
essays. They may be more challenging because you
are required to be brief yet complete. And after
you have answered these, try making up some
similar questions and then answering them.
Suggestion:
The Study Questions at the end of each text
chapter make very good short-answer essay
questions.

1. How can choice be **both** pressure and freedom?

2. How do age expectations affect choices people
    make regarding their personal relationships?

3. Explain clearly the life spiral concept and
    give an example of it.

<u>Essay</u>

The following are sample essay questions--
questions of the type you may be asked if your
instructor uses essay questions. Even if your
instructor does not use essay questions, you can
help organize and consolidate your learning if you
can answer these questions in a <u>well-organized</u>
and <u>complete</u> manner. Usually, the third essay
question is the most challenging.

1. According to the text, in what ways are
    personal choices affected by social influences
    such as historical events, race and ethnicity,
    social class, and age expectations? Be sure
    to support your answer with data or facts from
    the text.

2. In a well-organized essay, explore the idea of
    change as a usual rather than as an unusual
    aspect of family life. Support your answer
    with material from the text.

3. Karl Marx said that it is true that people
    make history, but they do not do it
    independently or under conditions of their own
    choosing. To what extent is this true about
    the history of individuals, couples, and
    families?

Answers

Chapter Summary

age expectations   family life **cycle**
historical events   choosing by **default**
they are **similar**   meant by the **life** **spiral**

Completion

1. age expectations
2. choosing by default
3. choosing
   knowledgeably
4. life spiral
5. familism
6. archival family
   function
7. individualistic
   (self-fulfillment)
   values
8. life spiral
9. family life cycle
   and its stages

True-False

1. T
2. F
3. F
4. T
5. T
6. T
7. F
8. F
9. T
10. T
11. F
12. F
13. T
14. T
15. F

Multiple Choice

1. d
2. b
3. a
4. d
5. b
6. d
7. b
8. a
9. a
10. c
11. b
12. b
13. d
14. b
15. d
16. d
17. c
18. c
19. d
20. c
21. b
22. c
23. b
24. a
25. d

CHAPTER 1
Exploring the Family

Chapter Overview

Chapter 1 explores the meaning of family, what it means to be a family member, and the role of families in the larger society. Because the family can be seen from different points of view or perspectives, the chapter reviews a variety of these points of view: family development theory, structure-functional theory, interactional theory, exchange theory, systems theory, conflict theory, and postmodern theories about family. Structure-functional theory is explored in detail in terms of the extent to which it is helpful in understanding families and family living in our society. The chapter concludes by reviewing the various methods by which family may be productively studied.

Chapter Summary

Chapter 1 provides an overview of marriages and families today and presents the challenge of defining "family." Is this close, face-to-face relationship--coined by Cooley a _____ group-- declining? Or is it changing? There is evidence of change, but is change necessarily decline? We live in a changing society, characterized by increasing economic, ethnic, and family structural diversity. As the text demonstrates, family diversity has progressed to the point that there is no "typical" family form today.

The family can be looked at from different points of view or perspectives: family development, structure-_____, interactional, systems, exchange, and conflict theories. These ways of looking at family help us to understand the many facets of what "family" means.

The family development perspective concentrates on how families _____ over time, sensitizing us to important family transitions and challenges. But its usefulness is somewhat constrained by this theory's assumption that families have a common trajectory or life course.

A view of the family as a social institution, whose values, norms, and activities are directed toward the performance of certain functions for society and for its members, is presented by the _____-functional theory of family. This perspective derives much of its power from cross-cultural and historical comparisons, but has been criticized for oversimplification and for neglecting conflict in family living.

A perspective that concentrates on day-to-day interpersonal communications and relationships between family members, whereby family emerges from the relationships, is the _____ theory of family. In this perspective, family interaction is seen as central to the process whereby partners as individuals and as family members define themselves for themselves and others. But interactional theory tends to minimize the importance of the family's external environment and of conflict in the family.

Bargaining and resources are key concepts in the _____ theory of family relationships. In this perspective, costs and rewards, resources and bargaining, equal and unequal exchanges are the focus of attention. But viewing the family as a complete network of relationships--like a large organic mechanism or a large organization--is the key identifying hallmark of _____ theory of family.

Looking at the opposite side of family cooperation and agreement, and emphasizing the role of power differences, is the _____ theory of family relationships. Some people have more power than others. Power inside the family is often related to power outside of the family. It is not uncommon for feminist and marxist sociologists to emphasize the illuminating power of this family perspective.

Discussing family from the perspective of postmodern society, separate rather than shared social worlds and diverse meanings of family are characteristic of recent _____ theories of family.

Social scientists have devised a number of research methods to obtain information about and to help understand family relationships. In

surveys, face-to-face or telephone interviews and
questionnaires are used to gather information.
Laboratory observations and _____ use
carefully controlled conditions to study family
relationships. Naturalistic _____ is
used to study family relationships in everyday,
natural settings, where these relationships
usually occur, as they occur spontaneously.
Clinicians' case studies study fewer instances of
individual partners and families, but these case
studies compiled by psychologists, marriage
counselors, social workers and others can offer
valuable insights into the realities experienced
by family members. Historical and cross-cultural
studies enrich our understanding of family
relationships by broadening our perspective and
testing our generalizations on other societies and
historical periods.
    By using a variety of theoretical
perspectives and research methods--not just one
theory or method--scientists can gain increasingly
complete knowledge about family living.

Key Terms

You should be able to explain the concepts listed
below. In your explanation, try to avoid using
the concept you are explaining. You should be
able to give several examples of each concept and
to explain why each example is an example.

primary group
  (p. 22)
family
  (p. 26)
family development
  theory
  (p. 28)
structure-functional
  theory
  (p. 28)
interactional theory
  (p. 30)

extended family
  (p. 38)
monogamy
  (p. 38)
polygamy
  (p. 38)
surveys
  (p. 42)
experiment
  (p. 43)
naturalistic
  observation
  (p. 44)

exchange theory
   (p. 31)
systems theory
   (p. 32
conflict theory
   (p. 32)
postmodern theories
   (p. 33
functional requisites
   (p. 34
social
   institutions
   (p. 34)
nuclear family
   (p. 38)

case studies
   (p. 44)
longitudinal studies
   (p. 45)
experiential reality
   (p. 41)
agreement reality
   (p. 41)
scientific
   investigation
   (p. 42)

Completion

Complete the following sentences by selecting the
correct alternative from the terms listed above.
Some may be used more than once. Some may not be
used at all. Filling in a blank may require more
than one word.

1. _____ refers to
   a type of marriage in which there is a
   sexually exclusive union of one man and one
   woman.

2. In _____, a person has
   more than one spouse.

3. In a(n) _____, people
   are involved in intimate, face-to-face
   relationships, communicate with one another
   as whole human beings, and give one another
   the feeling of being accepted and liked for
   what one is.

4. What members of a society do believe is true,
   or, _____, may
   misrepresent the actual realities experienced
   by families.

5. A _____ is any sexually expressive or parent-child relationship in which people live together with commitment in an intimate relationship and see themselves as attached to the group that has an identity of its own.

6. The theoretical perspective that stresses costs and rewards, resources and "deals" in relationships is _____.

7. The perspective that sees the family as a large network of connected parts, including all aspects of daily living, each affecting and in turn affected by the others, is called _____.

8. The concept of _____ refers to those tasks that must be performed if the group or society is to survive and maintain its identity.

9. The procedure called _____ uses various methods such as surveys, experiments, and case studies to come to reliable and valid conclusions.

10. In the research method called a(n) _____ the researcher asks questions of a representative sample of a larger group of people. The data gathered in this way are commonly analyzed using computers, and the social scientists then draw conclusions from these results.

11. The type of scientific investigation called a(n) _____ requires careful control and monitoring of almost every aspect of the research project. The conditions under which the research takes place are carefully controlled.

12. In using _____ the researcher records information gained by viewing the everyday "normal" behavior of the research subjects.

13. In the research procedure called
_____, the
researcher typically studies relatively few
people but gathers a great deal of
information about each one studied, providing
much vivid detail and realistic flavor of the
individuals and their daily settings.

14. Scientists conduct _____
when they get information about individuals,
families, or larger groups and do so in such
a way as to make comparisons over a long
period of time.

Key Research Studies

U.S. Bureau of the Census: Some facts about
today's marriages and families. While not a
single research study, the U.S. Bureau of the
Census continually collects and publishes
information about the U.S. population and is a
major source of data on changes in marriage and
family trends. Ask your librarian to show you
where to find these and similar reports in your
library.

Key Theories

family development theory   exchange theory
structure-functional theory   systems theory
interactional theory   conflict theory

True-False

____   1. The median age at first marriage in 1990
            is comparable to what it was at the turn
            of the century.

____   2. In recent years, people in the United
            States have been marrying earlier than
            in the period before 1980.

____ 3. Out-of-wedlock childbearing in this country is now more concentrated among women in their 20s than among teenagers.

____ 4. More and more people in the United States are now living alone than was the case in the past.

____ 5. Young adults in this country are increasingly likely to be living with their parents.

____ 6. The proportion of the population over age 65 has increased since 1960--by almost 40 percent.

____ 7. We can expect bonds between adult children and their parents, as well as between grown sisters and brothers, to become more important, not less so.

____ 8. Anglo women have about the same fertility rate as do minority women.

____ 9. Poverty is unevenly distributed across racial/ethnic groups.

____ 10. The proportion of the population below the poverty line has remained approximately steady since 1988.

____ 11. Approximately 28 percent of Latinos live below the poverty line.

____ 12. The legal definition of family has become much more flexible and non-specific.

____ 13. The structure-functional theory of family emphasizes developmental change in families.

____ 14. Structure-functional theory of family emphasizes what the family institution does for family members and for society.

_____ 15. The largest percentage of American households consists of married-couple families with children under 18.

_____ 16. Interactionist theories of family are interested in self and identity.

_____ 17. Viewing decision making within marriages as affected by the relative resources of the spouses characterizes exchange theory.

_____ 18. Systems theory emphasizes the effect of social structure on family systems.

_____ 19. A "survey" is not one of the several methods of investigation that can be called "scientific."

_____ 20. One of the disadvantages of surveys is that respondents tend to say what they think they should say rather than what they actually believe.

_____ 21. Scientists should use the experimental method when they are not in control of many aspects of the research projects.

_____ 22. Naturalistic observation is ordinarily not considered a form of scientific investigation because it is too informal.

_____ 23. A researcher selects three families that are believed typical of a certain type of family and reports in great detail about each family. This research procedure illustrates the case study method of investigation.

_____ 24. Longitudinal studies are those methods that gather data about research participants over a considerable period of time.

_____ 25. The strengths of one research tool can make up for the weaknesses of another.

1. The text states that people have been _____ marriage in recent years.
    a. avoiding
    b. "torpedoing"
    c. postponing
    d. "passive-alienating"

2. Currently about _____ of American women marry at some time during their lives.
    a. 50 percent
    b. 65 percent
    c. 82 percent
    d. 94 percent

3. In 1988, about _____ of all marriages were remarriages for one or both partners.
    a. 17 percent
    b. 24 percent
    c. 33 percent
    d. 46 percent

4. Overall, fertility has _____ since the 1950s and early 1960s.
    a. increased slightly
    b. decreased slightly and stayed at that level
    c. decreased dramatically, but has recently risen again
    d. not changed significantly

5. About how many families today contain no children under the age of 18?
    a. about half
    b. about three-quarters
    c. about one-third
    d. about one-quarter

6. The rate of divorce has more than doubled since 1965, peaking in _____ and dropping slightly since then.
    a. 1965
    b. 1979
    c. 1985
    d. 1990

7. As a consequence of divorce and unmarried parenthood, about _____ of all U.S. children under 18 lived with just one parent in 1990.
   a. one-sixth
   b. one-third
   c. one-fourth
   d. one-half

8. The labor force participation rate for unmarried mothers is _____ than that of married mothers.
   a. slightly higher
   b. dramatically higher
   c. lower
   d. no different

9. About _____ of all households in 1991 contained only one person.
   a. 13 percent
   b. 25 percent
   c. 37 percent
   d. 49 percent

10. The proportion of the population over 65 has increased by almost _____ since 1960.
   a. 15 percent
   b. 22 percent
   c. 31 percent
   d. 40 percent

11. Ten percent of whites live below the poverty line, compared with _____ of African-Americans and _____ of Latinos.
   a. 51 percent; 33 percent
   b. 31 percent; 16 percent
   c. 40 percent; 20 percent
   d. 64 percent; 20 percent

12. Andrea and Paul--and their two preteen children--spend time together, share feelings, are often proud of one another, are sometimes angry at one another, and have deep feelings about the value of their relationships. They illustrate:
   a. primary group
   b. extended family
   c. value cleavage
   d. polygamy

13. A _____ is characterized by close face-to-face relationships in which people communicate with one another as whole human beings, and people maintain these relationships because they enjoy the relationship for its own sake.
    a. normative group
    b. tertiary group
    c. bureaucratic group
    d. primary group

14. Which of the following is not one of the elements of the text's definition of "family"?
    a. legitimated by some civil or religious ceremony
    b. live together with commitment
    c. form an economic unit and care for any young
    d. find their identity as importantly attached to the group

15. The theory that concentrates on how families change over time and on how families change as time goes by is:
    a. structure-functional theory
    b. exchange theory
    c. conflict theory
    d. family development theory

16. The strength of _____ theory is in the comparative study of society, sensitizing us to differences between various societies at various times.
    a. structure-functional
    b. conflict
    c. exchange
    d. interactional

17. According to Figure 1.1, The many kinds of American households, 1991, the largest percent of households is in which category?
    a. married-couple families with children under 18
    b. male- or female-headed family households
    c. child-free or post-child-rearing marriages
    d. people living alone

18. The view that divorce is the dismantling of the couple relationship, with newer identities arising that are not tied to the marriage, typifies _____ theories of family.
   a. exchange
   b. conflict
   c. interactional
   d. structure-functional

19. Exchange theory has as its focus the _____ of the partners.
   a. conflict competency
   b. differential resources
   c. networking ability
   d. interactional

20. Family interaction based on unequal power is a hallmark of _____ theory.
   a. structure-functional
   b. exchange
   c. postmodern
   d. conflict

21. Betty and Fred remained married as long as they were both living. They had two children who lived with them until both children finished college. They grew old together, and when Betty died, Fred never remarried. This illustrates:
   a. polygamy
   b. monogamy
   c. family linkage
   d. fictive kin

22. About ____ of the world's cultures insist on monogamous marriage.
   a. 20 percent
   b. 45 percent
   c. 70 percent
   d. 90 percent

23. Three or more generations with social relationships and economic exchanges between the generations is called a(n):
   a. nontraditional family
   b. traditional family
   c. extended family
   d. cooperative-utilitarian family

24. In the United States, the belief that members of a family almost always reside together is an example of:
    a. an inaccuracy caused by "blinders"
    b. a statistical fact
    c. a statistical generalization
    d. a reflexive institution

25. Sally studied conflict by randomly assigning her student research subjects to two groups. She gave both groups the same family conflict problem to resolve, but she gave one group a mini-lecture on conflict resolution. Then she measured how long it took each group to resolve its family conflict problem. This is an example of what type of investigation?
    a. survey
    b. experiment
    c. naturalistic observation
    d. clinician's case study

26. Hector programed the computer to select at random the names of 500 students, who were then individually asked the same questions by trained interviewers. Hector then used the computer to analyze the results. Hector used which type of scientific investigation?
    a. the longitudinal study
    b. a survey
    c. the case study
    e. experiments

> **Your Opinion, Please:** Defining the family isn't easy. Some sociologists think that a husband, wife, and two children living together is <u>very</u> different from a husband and wife who don't have any children and never have any-- so different, in fact, that some want to use different concepts to refer to the two situations. Some want to call the husband, wife, and two children a "family" and call the husband and wife with no children a "married pair." What is your opinion? Do you think a permanently child-free couple is a family? Or do you think that a different term should be used for such a situation? **Why** do you think so?

Short-Answer Essay Questions

The following are sample short-answer essay questions--questions of the type you may be asked if your instructor uses questions like these. Even if your instructor does not use questions like these, you can help organize and consolidate your learning if you can answer these questions in a <u>well-organized</u> and <u>complete</u> manner. Do not think that brief essays are easier than lengthy essays. They may be more challenging because you are required to be brief yet complete. And after you have answered these, try making up some similar questions and then answering them. The Study Questions at the end of each text chapter make very good short-answer essay questions.

1. Distinguish between family development theory and interactional theory.

2. Compare and contrast the conflict theory with exchange theory. In what ways are they both similar and different?

3. Explain how personal experience can act as "blinders," preventing more complete understanding of couples and families.

Essay

The following are sample essay questions-- questions of the type you may be asked if your instructor uses essay questions. Even if your instructor does not use essay questions, you can help organize and consolidate your learning if you can answer these questions in a well-organized and complete manner. Usually, the third essay question is the most challenging.

1. Explain family development and postmodern theory of family. In what ways are they similar? In what ways are they different?

2. None of the various theories of family is "the correct theory." Explain why.

3. Explore the strengths and weaknesses of each of the five methods of social research.

Answers

Chapter Summary

by Cooley a **primary** group
structure-**functional**
how families **change** over time
by the **structure**-functional theory
is the **interactional** theory
in the **exchange** theory of family
hallmark of **systems** theory
is the **conflict** theory
of recent **postmodern** theories
observations and **experiments**
naturalistic **observation**

Completion

1. monogamy
2. polygamy
3. primary group
4. agreement reality
5. family
6. exchange theory
7. systems theory
8. functional requisites

9. scientific
    investigation
10. survey
11. experiment
12. naturalistic
    observation
13. case studies
14. longitudinal
    studies

True-False

1. T
2. F
3. T
4. T
5. T
6. T
7. T
8. F
9. T
10. F
11. F
12. T
13. F

14. T
15. F
16. T
17. T
18. F
19. F
20. T
21. F
22. F
23. T
24. T
25. T

Multiple Choice

1.	c	14.	a
2.	d	15.	d
3.	d	16.	a
4.	c	17.	c
5.	a	18.	c
6.	b	19.	b
7.	c	20.	d
8.	c	21.	b
9.	b	22.	a
10.	d	23.	c
11.	b	24.	a
12.	a	25.	b
13.	d	26.	b

CHAPTER 2
Gender Stereotypes, Scripts,
and Stratification

Chapter Overview

This chapter explores gender stereotypes.
The text explores gender identity arguments based
on genetics, based on social influences, and on
the interaction on biology and society. The role
of socialization is explored, with attention given
to the power of cultural images, family
socialization, and school. Finally, gender in
adulthood is reviewed with particular attention to
the role of gender regarding stress, personal
change, social movements, and related issues such
as androgyny and ambivalence about gender.

Chapter Summary

Roles of men and women have changed over time, but
living in our society remains a different
experience for women and for men. Gender
_____ and scripts influence people's
behavior, attitudes, and options. Women tend to
be seen as more expressive, relationship-oriented,
and "communal"; men are considered more
instrumental, or _____. Stereotypes of African-
American men and women are more similar to each
other than are those of other Americans.
Generally, traditional masculine scripts require
that men be confident and self-reliant,
occupationally successful and engage in "no sissy
stuff." During the 1980s the "new male" (or
"liberated male") script emerged, according to
which men are expected to value tenderness and
equal relationships with women. Traditional
_____ involve a woman's being a man's
helpmate and a "good mother." An emergent
feminine script is the successful "professional
woman"; when coupled with the more traditional
scripts, this results in the _____ script.
    The extent to which actual men and women
differ from one another and follow these
stereotypes and scripts can be visualized as two
overlapping normal distribution curves. Means
differ according to cultural expectations, but

within-group variation is usually greater than
_____ variation. An exception is male
dominance, evident in politics, religion, and in
the economy.

Biology interacts with culture to produce
human behavior, and the two influences are
difficult to separate. Sociologists give greater
attention to the _____ process, for which
there are several theoretical explanations.
Overall, however, sociologists stress
socialization in the family, in play and games,
and in school as encouraging gendered stereotypes,
scripts, and behavior.

Women and men make choices in a context of
changing expectations at work and in
relationships. Change brings mixed responses,
depending on class, racial/ethnic groups,
religion, or other social indicators. New
cultural ideals are far from realization, and
efforts to create lives balancing love and work
involve conflict and struggle. But the greater
equality that allows conflict between partners
also creates the conditions for satisfying
intimate relationships.

Point to Ponder: If female students and male
                  students have different
achievement levels in computer science courses
--and there is evidence that they do--is it
because male students and female students have
different socialization experiences in terms
of learning computer science? One study found
that when female students in computer science
had difficulties, teaching assistants tended to
"fix" their problems for them rather than
explain computer principles and let them "fix"
their own problems, possibly learning more in
the process. If male and female computer
science students are found to have different
ability levels in computer science at the end
of a course with this type of instruction, do
you think it is because of innate or "natural"
differences between males and females, or do
you think it might be more likely related to
differences in learning experiences?   Can
you think of any other kinds of academic
differences that could be examples of similar
socialization experiences?

Key Terms

You should be able to explain the concepts listed
below. In your explanation, try to avoid using
the concept you are explaining. You should be
able to give several examples of each concept and
to explain why each example is an example.

sex
  (p. 51)
gender role
  (p. 51)
gender
  (p. 51)

assignment
  (p. 59)
internalization
  (p. 60)
socialization
  (p. 60)

instrumental character
   traits
   (p. 52)
expressive character
   traits
   (p. 52)
hormone
   (p. 59)
hermaphrodite
   (p. 59)

social learning theory
   (p. 62)
self-identification
   theory
   (p. 62)
Chodorow's theory
   (p. 62)
androgyny
   (p. 76)
intimacy
   (p. 78)

Completion

Complete the following sentences by selecting the
correct alternative from the terms listed above.
Some may be used more than once. Some may not be
used at all. Filling in a blank may require more
than one word.

1. Biological and/or physiological differences
   between males and females are referred to as
   _____ differences.

2. The ways in which men are traditionally
   expected to behave and in which women are
   traditionally expected to behave are called
   _____.

3. _____ does not refer to
   biological characteristics; it refers to the
   psychological, cultural, or social
   consequences of biological characteristics.

4. Competitiveness, self-confidence, logic, and
   nonemotionality as allegedly "natural" for
   men are examples of _____.

5. _____ are perceived to
   be more "feminine," and include sensitivity
   to the needs of others and the ability to
   express tender feelings.

6. Sex _____ influence sexual dimorphism:
   sex-related differences in body structure and
   size, muscle development, and voice quality.

7. Studies of hermaphrodites point out the importance of _____ in determining how males and/or females behave.

8. _____ is the process whereby people learn to make social and cultural attitudes a part of themselves, a part of who and what they are as persons.

9. The general process by which society influences people to internalize attitudes and expectations is called _____. Because of this, people usually exhibit the kinds of behavior people expect of them. They have, after all, been "raised that way."

10. The term _____ refers to the social and psychological condition in which individuals can think, feel, and behave both instrumentally and expressively, showing both masculine and feminine characteristics.

11. Paying little or no attention to sexual differences as far as role behavior is concerned but instead expecting behavior to be appropriate to either sex is called _____.

12. Inexpressiveness is often a major barrier to _____ in marriage.

_____

Key Research Studies

    Money and Ehrardt: conclusions based on
        studies about hermaphrodites
    Chodorow: theory of gender
    Maccoby and Jacklin: innate vs. learned
        gender differences

Socialization and gender
Chodorow's theory of gender
Androgyny and egalitarianism

True-False

____ 1. Gender influences perhaps as much as three-fourths of a person's life and relationships.

____ 2. Gender is a central organizing principle of all societies.

____ 3. In the context of this chapter, "sex" refers only to male or female physiology.

____ 4. "Gender" involves social roles and stereotypical cultural scripts regarding what it means to be masculine or feminine.

____ 5. Almost all people internalize gender-related attitudes and expectations for behavior.

____ 6. "Instrumental character traits" are those that include such traits as logic, confidence, and competitiveness.

____ 7. Differences among women (or among men) are usually greater than differences between men and women.

____ 8. In groups, males tend to dominate verbally.

____ 9. Ethologists are scientists who specialize in the study of the ethical systems of various cultures, including the ethics of family law.

____ 10. Hermaphrodites, named for the Greek god Hermes and the goddess Aphrodite, are persons who depend too much on satisfactions they receive from relationships with the opposite sex.

____ 11. The only difference between boys and girls that Maccoby and Jacklin found to be biologically based was "aggressiveness."

____ 12. To have an androgynous marriage is to have a marriage in which male and female roles are highly traditional.

____ 13. "Androgyny" and "hermaphrodite" are not interchangeable concepts.

____ 14. Men as well as women experience ambivalence in regard to women's changing roles.

____ 15. Haas's study of role-sharing marriages found that couples tended to adopt role sharing for ideological reasons, and not out of dissatisfaction with traditional role allocations.

____ 16. Haas's study of role sharing marriages indicates that this kind of role sharing is basically unworkable, that it tends to create more problems than it solves.

____ 17. Traditional sex roles are hazardous to women's health.

____ 18. A study of role-sharing couples provides evidence of the mainly negative consequences of androgyny and egalitarianism.

____ 19. In a study of role-sharing couples, females showed less resentment of the male partner's power in the relationship.

1. Chapter 2 on gender roles is most directly
    and immediately concerned with which of the
    following controversies?
    a. the group versus the individual
    b. the society versus the culture
    c. technology versus the individual
    d. heredity versus social environment

2. According to your text, _____ refers to
    attitudes and behavior associated with and
    expected of the two sexes.
    a. gender
    b. archetypes
    c. genotypes
    d. sexual dimorphism

3. _____ are culturally written and
    directed "plots" for human behavior.
    a. gender maps
    b. cultural scripts
    c. socialization agendas
    d. normative trajectories

4. According to your text, _____ study human
    beings as an evolved animal species just like
    any other evolved animal species.
    a. anthropologists
    b. ethnomethodologists
    c. ethologists
    d. structuralists

5. Women have always been involved in:
    a. actively pursuing civil rights
    b. economic production
    c. sex-differentiated political behavior
    d. social control through manipulation of
        religion

6. Studies such as those of hermaphrodites and
    "assignment" point out the connectedness of:
    a. the civil and personal worlds
    b. free will and bureaucratic demands
    c. psychology versus sociology
    d. genetics and society

7. Your text argues that human biological realities and sociocultural realities:
   a. interact with each other
   b. are in contradiction with each other
   c. are involved with each other in an additive manner
   d. neutralize each other

8. Research shows that _____ is important to _____.
   a. sex-role interchangeability; the Pawnee
   b. a sense of sharing; narcissists
   c. assignment; hermaphrodites
   d. biological determinism; ethnomethodologists

9. According to the text, the only difference between boys and girls that seems to be biologically based is:
   a. aggressiveness
   b. competitiveness
   c. math ability
   d. hand/eye coordination

10. Which of the following has been investigated as it relates to gender-role identity among females?
   a. testosterone concentration levels
   b. the release of pitocin during interaction with males
   c. brain-stem response type during blood monitoring
   d. dilation of the iris of the eye

11. The text states that _____ accentuates male-female differences rather than similarities.
   a. economics
   b. religion
   c. language
   d. anthropology

12. The process by which society influences members to internalize attitudes and expectations for behavior is called:
   a. institutionalization
   b. parasocial relationships
   c. formal education
   d. socialization

13. Chodorow suggests that _____ emotional life is richer and the sense of self less problematic, as compared with that of_____.
   a. girls'; boys
   b. middle-aged men's; middle-aged women
   c. straights'; gays
   d. Hispanics'; Asians

14. According to Chodorow's theory of gender, the task of _____ is full of stress and anxiety for both sexes.
   a. sexual gratification
   b. emotional focusing
   c. intimacy
   d. separation

15. The correct answer in the above question, according to the text, is MOST difficult for:
   a. heterosexual mothers
   b. bisexual mothers
   c. daughters
   d. sons

16. Chodorow's theory of gender helps explain why men develop _____ character traits and women develop _____ ones.
   a. sexual; erotic
   b. sex-role; gender-role
   c. more positive; more neutral
   d. instrumental; expressive

17. Among blacks, gender stereotypes are:
   a. less strong than among whites
   b. much stronger for males than for females
   c. much stronger for females than for males
   d. stronger among younger than among older blacks

18. Raising sons and daughters differently is called:
    a. socialization cleavage
    b. differential socialization
    c. cultural discontinuity
    d. subcultural socialization

19. Research has shown that when a child's older sibling _____, play activities tend to be gender stereotyped.
    a. has child-care responsibilities
    b. is of the same sex
    c. has a troubled school experience
    d. is an adopted child

20. Boys more often than girls:
    a. play games in which there is neither loser nor winner
    b. single out a scapegoat or source of blame when the team does not play well
    c. play in fairly large groups with a hierarchical "pecking order"
    d. spend quality time in game preparation and in analyzing the failure of the team to perform well in game situations

21. Research shows that, at least among white children, _____ pay more attention to males than to females.
    a. parents
    b. mothers
    c. toy salespersons
    d. teachers

22. Female faculty members serve as role models for female college students. However, about _____ of college teachers are women.
    a. 64 percent
    b. 42 percent
    c. 28 percent
    d. 14 percent

23.   Data suggest that college students are:
      a.   more gender-typed than persons who do not
           attend college
      b.   about as gender-typed as are their
           parents
      c.   more gender-typed when they finished
           college than when they started college
      d.   more concerned with gender than is the
           public in general

24.   The text takes the view that traditional
      gender expectations can be seen as
      _____ for women and men in our
      culture.
      a.   mandatory
      b.   limiting
      c.   expansive
      d.   largely optional

Your Opinion, Please: Some people think that
                          it works out best for
all concerned if "men are men" and "women are
women," with each sex displaying gender roles
that most people in society expect. But some
people think that such notions are too narrow
and keep people from having new experiences,
from exploring new innate preferences. What
are your views on this issue? Are you in
support of a world in which gender roles are
highly predictable or would you prefer a world
in which people are more androgynous?

Short-Answer Essay Questions

The following are sample short-answer essay
questions--questions of the type you may be asked
if your instructor uses questions like these.
Even if your instructor does not use questions
like these, you can help organize and consolidate
your learning if you can answer these questions in
a well-organized and complete manner. Do not
think that brief essays are easier than lengthy

essays. They may be more challenging because you are required to be brief yet complete. And after you have answered these, try making up some similar questions and then answering them. Suggestion: The Study Questions at the end of each text chapter make very good short-answer essay questions.

1. To what extent do individual men and women fit gender stereotypes and scripts?

2. In what way--and to what extent--do language and the media convey gender stereotypes and scripts? Give a persuasive example of each phenomenon--language and media.

3. Summarize the relationship play and games have regarding socialization to gender roles.

Essay

The following are sample essay questions-- questions of the type you may be asked if your instructor uses essay questions. Even if your instructor does not use essay questions, you can help organize and consolidate your learning if you can answer these questions in a well-organized and complete manner. Usually, the third essay question is the most challenging.

1. Has gender equality been achieved in the economic sphere, or is there still need for change? Support your answer with facts.

2. In what ways does day-to-day life produce
    gender differences? Support your answer with
    data and specific research results.

3. There _are_ noticeable gender differences in
    society. Use what you have learned in this
    chapter to explore the positive and negative
    consequences of gender differences. In
    answering this question, there is room for
    your opinion--after you have set forth "the
    facts."

Answers

Chapter Summary

gender **stereotypes**
instrumental, or **agentic**
traditional **feminine scripts**
results in the **superwoman** script
greater than **between-group** variation
attention to the **socialization** process

Completion

1.	sex	8.	internalization
2.	gender roles	9.	socialization
3.	gender	10.	androgyny
4.	instrumental character traits	11.	androgyny
5.	expressive traits	12.	intimacy
6.	hormones		
7.	assignment		

True-False

1.	F	11.	T
2.	T	12.	F
3.	T	13.	T
4.	T	14.	T
5.	T	15.	F
6.	T	16.	F
7.	T	17.	T
8.	T	18.	T
9.	F	19.	T
10.	F		

Multiple Choice

1.	d	13.	a
2.	a	14.	d
3.	b	15.	d
4.	c	16.	d
5.	b	17.	a
6.	d	18.	b
7.	a	19.	b
8.	c	20.	c
9.	a	21.	d
10.	a	22.	c
11.	c	23.	c
12.	d	24.	b

CHAPTER 3
Loving Ourselves and Others

Chapter Overview

Chapter 3 examines love as an aspect of personal, cultural, and social life. It begins by exploring what it is about society today that makes love so important to so many people. The chapter then investigates what love _is_ and what it _isn't._ Love is seen as a phenomenon with its own requirements in terms of the characteristics of the individuals involved. Self-esteem is revealed as a prerequisite to loving. The chapter closes by explaining that love is not a fact but a process-- a process of discovery.

Chapter Summary

In an impersonal society, _love_ provides an important source of fulfillment and intimacy. Genuine loving in our society is rare and difficult to learn.   Our culture's emphasis on self-reliance as a central virtue ignores the fact that all of us are _____, relying on parents, spouses or partners, other relatives, and friends far more than our culture encourages us to recognize. Loving is one form of interdependence.
    In the _triangular theory of love,_ Sternberg uses the three dimensions of intimacy, passion, and commitment to generate a typology of love, one of which--_____ love --involves all three components. John Lee lists six love styles: _eros,_ or passionate love; _storge,_ or familiar love; _pragma,_ or pragmatic love; _____, or altruistic love; _ludus,_ or love play; and _____, or possessive love.
    Despite its importance, however, love is often misunderstood. It should not be confused with _martyring_ or _____.   There are many contemporary love styles that indicate the range of dimensions love-like relationships--not necessarily love--can take.
    Self-esteem seems to be a prerequisite for loving. In relationships where there is emotional _interdependence,_ love probably involves acceptance

of oneself and others, a sense of empathy, and a
willingness to let down barriers set up for self-
preservation. A-frame, H-frame, and ____-frame
relationships are some of the forms love can take.
People discover love; they don't simply find
it. Put differently, love is not an event; it is
an unfolding process. Reiss's _____ of love
sets forth four basic stages of the love process:
rapport, _____, mutual dependency,
and personal need fulfillment.

     Since partners need to keep on sharing their
thoughts, feelings, troubles, and joys with each
other, love is a continual process.

Key Terms

You should be able to explain the concepts listed
below. In your explanation, try to avoid using
the concept you are explaining. You should be
able to give several examples of each concept and
to explain why each example is an example.

love
   (p. 86)
emotion
   (p. 86)
legitimate needs
   (p. 87)
illegitimate needs
   (p. 87)
triangular theory of love
   (p. 92)
intimacy
   (p. 92)
passion
   (p. 93)
commitment
   (p. 93)
consummate Love
   (p. 93)
love styles
   (p. 94)
eros
   (p. 94)
storge
   (p. 94)

ludus
   (p. 95)
mania
   (p. 95)
martyring
   (p. 96)
manipulating
   (p. 96)
symbiotic relationships
   (p. 97)
self-esteem
   (p. 97)
narcissism
   (p. 97)
dependence
   (p. 99)
independence
   (p. 99)
interdependence
   (p. 99)
A-frame relationship
   (p. 100)
H-frame relationship
   (p. 100)

pragma
   (p. 95)
agape
   (p. 95)

M-frame relationship
   (p. 100)
wheel of love
   (p. 102)
self-disclosure
   (p. 102)

Point to Ponder: Perhaps most people in our society think that romantic love is the appropriate basis for marriage. And yet we know that not all societies base marriage on individual choice. Some societies have "arranged marriage." In such societies, parents, siblings, or hired match-makers arrange to pair people up with each other. For the couple, love is largely left to later after the marriage. Think about the possible positive and negative consequences of such arranged marriages. What are the positive and negative consequences, from your point of view? Do you think that the positive consequences outweigh the negative ones? Try to imagine yourself in an arranged marriage. Do you like what you see? Why? Or why not?

Completion

Complete the following sentences by selecting the correct alternative from the terms listed above. Some may be used more than once. Some may not be used at all. Filling in a blank may require more than one word.

1. A(n) _____ is a strong feeling, arising without conscious mental or rational effort, that motivates an individual to behave in a certain way.

2. _____ are those arising in the present rather than out of deficits or failures accumulated in the past.

3. _____ spring from feelings of self-doubt, unworthiness, and inadequacy.

4. According to Sternberg's triangular theory of love, when intimacy, passion, and commitment are all involved, the result is the type of love called _____.

5. _____ refer to the various distinctive "personalities" that love-like relationships can take.

6. The type of love that is characterized by intense emotional attachment and powerful sexual feelings or desires is termed _____.

7. The term _____ refers to an affectionate, companionate style of loving.

8. _____ is the kind of love involving an emphasis on the practical side of human relationships, emphasizing economic and emotional security.

9. Often called altruistic love, _____ emphasizes unselfish concern for the beloved.

10. The type of love that emphasizes playfulness and the humorous or amusing aspects of the love relationship is what is meant by _____.

11. An insatiable need for attention and affection alternating between euphoria and depression is characteristic of a _____ type of love style.

12. _____ involves maintaining relationships by giving others more than one receives in return, usually with good intentions, but seldom with the feeling that genuine affection is being received.

13. _____ means seeking to control the feelings, attitudes, and behavior of the partner by subtle, indirect ways rather than by straightforwardly stating one's case.

14. Martyrs and manipulators are often attracted to each other, forming what counselor John Crosby calls a(n) _____.

15. _____ refers to an evaluation a person makes and maintains of her- or himself that expresses attitudes of approval or disapproval, success or failure, worth or unworthiness, and similar ideas.

16. _____ is essentially another word for selfishness.

17. The concept of _____ refers to a general reliance on another person or on other people for continual support or assurance, along with subordination.

18. _____ refers to self-reliance, self-sufficiency, sometimes including separation or isolation from others.

19. When we label a relationship _____, we imply that the people involved have high self-esteem and make strong commitments to each other.

20. In _____ the partners have strong couple identity but little individual self-esteem.

21. In _____ relationships, the partners stand self-sufficient and virtually alone, with neither influenced much by the other.

22. In _____ each partner
    has high self-esteem, experiences loving as a
    deep emotion, and is involved in mutual
    influence and emotional support.

23. The stages of rapport, self-disclosure,
    mutual dependence, and personality need
    fulfillment characterize the
    _____.

24. The first stage in the wheel theory of the
    development of love is _____.

25. In Reiss's wheel theory of the development of
    love, _____ follows self-
    revelation and precedes personality need
    fulfillment.

Key Research Studies

Cancian: Do men and women "care"
  differently?
Sternberg: the triangular theory of love
Lee: six love styles

Key Theories

Sternberg: the triangular theory of love
Crosby: three types of interdependence
  relationships
Reiss: the wheel of love

True-False

____ 1. Modern society in the United States is
          often characterized as impersonal.

____ 2. The text seems to feel that people in
          modern society are overly oriented to
          self-sacrifice and concern for others.

____ 3. Romanticizing is an infrequent aspect of
          the falling in love experience.

____ 4. "Legitimate needs" are those that enjoy the approval of recognized social institutions, such as family, friends, and work associates.

____ 5. "Illegitimate needs" are those needed by lovers who were themselves born out of wedlock or to unmarried mothers.

____ 6. Intimacy, passion, and commitment are all part of consummate love in the triangular theory of love.

____ 7. Storge is a type of love that refers to the playful, comical elements of the relationship.

____ 8. The pragma type of love emphasizes the practical element in human relationships.

____ 9. Agape refers to a type of love typified by unselfish concern for the beloved.

____ 10. Denise can't stop thinking about Freddie, even for an hour. She has almost nothing else on her mind except trying to satisfy her intense emotional and sexual attraction to Freddie, who maddeningly remains just beyond her reach. This is an example of the love style called "mania."

____ 11. The text feels that martyring is one of the highest types of love relationships.

____ 12. Manipulating is an activity that is a normal, natural, and productive part of the love relationship, regardless of the general public's negative view of manipulation.

____ 13. "Manipulating" is not a type of love, but "martyring" is a type of love.

____ 14. Being in love with another person involves changing yourself to be more like what the other person wants you to become.

_____ 15. A-frame relationships are characterized by couples in which the individuals have a strong couple identity but low levels of self-esteem.

_____ 16. In H-frame relationships, the partners stand virtually alone and isolated, with little or no couple identity and little emotionality.

_____ 17. In Reiss's wheel theory of love, the first stage is "rapport."

_____ 18. The moment Frank met Annette, he shared with her some things about himself about which most people weren't aware. Annette and he then began feeling more comfortable with each other. Frank and Annette have reversed the first two stages in the wheel theory of love.

_____ 19. "Mutual dependence" is the fourth stage in Reiss's wheel theory of love.

Multiple Choice

1. One of the reasons love may be so important to so many people in the United States is because much of everyday life is:
   a. perplexing
   b. impersonal
   c. ambiguous
   d. unrewarding

2. Some who have studied modern society are warning against the _____ in our social world. The text seems to point to this issue as one of the reasons love is so important to people today.
   a. extreme rationality and individuality
   b. low levels of goal orientation and persistence
   c. low levels of logical planning and problem solving
   d. inordinately high levels of religious influence

3. The concept of illegitimate needs directly involves which of the following?
   a. failures from the past that remain "unfinished business"
   b. requests by the loved one that just can't be met by any reasonable person
   c. needs that have the potential to "snowball" into something more
   d. needs that most people would view as somewhat deviant

4. According to the text, men in our society have great difficulty in:
   a. bridging the gap between liking and loving
   b. falling in love with women of their own social class
   c. establishing relationships between their lovers and their own parents
   d. communicating about love verbally

5. Cancian argues that in our society _____ are made to feel primarily responsible for the endurance or success of love.
   a. parents who socialized the couple
   b. psychological and psychiatric counselors and therapists
   c. women
   d. "activity" industries, such as cruise boats, restaurants, and ski lodges

6. The case study about Sharon and Gary, who discover love after twenty-five years, is about a couple who:
   a. marry, separate, and then rediscover love
   b. discover love after twenty-five years of singlehood
   c. are--separately--widowed, but then discover that they are still capable of loving
   d. marry, divorce, and then each finds love with a different partner

7. Which of these is a friendly, affectionate, and companionate style of love?
   a. storge
   b. eros
   c. pragma
   d. agape

8. The playful aspect of love is what is meant
    by which of the following styles of love?
    a. eros
    b. mania
    c. pragma
    d. ludus

9. Mike and LaPrelle feel strong love for each
    other. Perhaps the most notable
    characteristic of their love relationship is
    that they have such a good time, get such a
    "kick" out of one another, and spontaneously
    laugh and joke as they go about their
    activities.   This is a hallmark of a(n)_____
    love relationship.
    a. storge
    b. eros
    c. agape
    d. ludus

10. Unselfish concern for the beloved is the
    basic idea in which of the following styles
    of love?
    a. eros
    b. pragma
    c. ludus
    d. agape

11. Moodiness, jealousy, and insatiable need are
    signs of which of the following love styles?
    a. ludus
    b. pragma
    c. mania
    d. agape

12. It is characteristic of martyrs that they
    think that:
    a. love is something worth dying for
    b. it is better to be wanted as a victim
        than not to be wanted at all
    c. there is nobility in making the beloved
        do what you want him/her to do
    d. one should not compromise in order to
        sustain a love relationship

13. In the chapter about loving ourselves and others, "manipulating" is best summarized as which of the following?
    a. "I understand what you feel. Now let me tell you how I feel." (said softly)
    b. "I see what you have done. This what I'm gonna do now." (openly stated)
    c. "If you will do what I want you to do, only then will I do what you want me to do." (hinted, but not said openly)
    d. "If you do not do as I wish, then I will do as I threatened." (openly stated)

14. Manipulators, like martyrs, do not believe that:
    a. love really exists; they are cynical about love
    b. loving is a normal, desirable phenomenon; they see it as undesirable
    c. the partner is worth loving; they seriously undervalue the partner
    d. they are lovable or that others can really love them

15. Family counselor John Crosby has discussed _____ in which each partner depends on the other for a sense of self-worth.
    a. noblesse oblige
    b. benign parasitism
    c. mutual dependence neurosis
    d. symbiotic relationships

16. The feelings people have about their own worth are called their feelings of:
    a. self-concept
    b. self-esteem
    c. personal value
    d. personal assessment

17. The text states that _____ is a prerequisite for loving.
    a. self-esteem
    b. narcissism
    c. androgyny
    d. objectively measured sexual expertise

18. The two members of a couple confirm that they are involved in a couple relationship. They feel it is satisfying. Basically, however, each member is self-sufficient and is in many ways isolated and independent from the other person in the relationship. This situation describes a(n):
    a. A-frame relationship
    b. H-frame relationship
    c. K-frame relationship
    d. M-frame relationship

19. A thorough reading of the text's discussion of "discovering" love is to see love viewed as which of the following:
    a. a process or sequence
    b. a rare but marvelous coincidence
    c. an objective fact
    d. an encounter with a unique kind of love

20. People _____ love; they don't just find it.
    a. invent
    b. prolong
    c. discover
    d. defend themselves against or thwart

21. The second stage in Reiss's wheel theory of love is:
    a. mutual dependence
    b. personality need fulfillment
    c. rapport
    d. self-revelation

22. Peeling away outer layers of the self, outer layers of "protective coloration" we use to shield our vulnerability, and in the process showing more and more of the "real" or authentic self, is part of the _____ stage in Reiss's wheel theory of love.
    a. first
    b. second
    c. third
    d. fourth

23. Bruce and Jane have some things they share in common but they also have some individual activities. The things they share are very important to them: they divide the activities of these tasks, enjoy these tasks most when the other person is there as well, and really would prefer to forgo the activity rather than do it without the other person being there to take part.   This is part of the _____ stage in Reiss's wheel theory of love.
    a. first
    b. second
    c. third
    d. fourth

24. Carolyn and Joel have come to feel that they need to be together always because each feels that the other person completes them and makes them a better person because of the relationship. This is part of the _____ stage in Reiss's wheel theory of love.
    a. first
    b. second
    c. third
    d. fourth

25. Lisette and Claude didn't feel comfortable with one another from the moment they were introduced. They just couldn't get through those first moments comfortably. According to Reiss's wheel theory of love, they were unable to establish _____ when they met.
    a. self-revelation
    b. mutual dependency
    c. rapport
    d. personality need fulfillment

Your Opinion, Please: How do you know when you are "in love" with someone? A few questions that come to mind are these: Can you be "in love" with a person and yet dislike him or her as a person? Is there more than one person in the world with whom you could be "in love"? Is it possible to be "in love" with more than one person at a time? After you have given your own opinion, think about what the authors of the text would answer to each of the above questions, based on what they have written in this chapter. If you are feeling bold, ask these same questions of someone with whom you have a relationship. Next, do you feel comfortable sharing with that person your own answers to the questions? Why or why not?

Short-Answer Essay Questions

The following are sample short-answer essay questions--questions of the type you may be asked if your instructor uses questions like these. Even if your instructor does not use questions like these, you can help organize and consolidate your learning if you can answer these questions in a _well-organized_ and _complete_ manner. Do not think that brief essays are easier than lengthy essays. They may be more challenging because you are required to be both brief yet complete. And after you have answered these, try making up some similar questions and then answering them. Suggestion: The Study Questions at the end of each text chapter are very good short-answer essay questions.

1. Distinguish between legitimate needs and illegitimate needs that love might satisfy.

2. What are the main parts of the triangular theory of love? How does consummate love fit into this theory? Explain carefully, briefly, but completely.

3. Sometimes lovers say they would be willing to "do anything" for the sake of their beloved. Is this the same thing as "martyring"? Why or why not?

4. In Reiss's wheel theory of love, why does the rapport stage precede the self-revelation stage? Explain in terms of process.

Essay

The following are sample essay questions-- questions of the type you may be asked if your instructor uses essay questions. Even if your instructor does not use essay questions, you can help organize and consolidate your learning if you can answer these questions in a well-organized and complete manner. Usually, the third essay question is the most challenging.

1. Explain what love is and what it is not.

2. In what way(s) is self-esteem a prerequisite to love or loving? That is, does self-esteem seem to be required before one can fully love?

3. Compare and contrast Sternberg's triangular theory of love with Reiss's wheel theory of love. Do they have similar goals? Do they address similar or different issues? Explain in a well-organized essay.

Answers

Chapter Summary

all of us are **interdependent**
one of which-- **consummate** love
agape, or altruistic love
mania, or possessive love

martyring or
   manipulating
M- frame
wheel of love
rapport, **self-revelation**

Completion

1. emotion
2. legitimate needs
3. illegitimate needs
4. consummate love
5. love styles
6. eros
7. storge
8. pragma
9. agape
10. ludus
11. mania
12. martyring
13. manipulating
14. symbiotic relationship
15. self-esteem
16. narcissism
17. dependence
18. independence
19. interdependent
20. A-frame
21. H-frame
22. M-frame
23. wheel of love
24. rapport
25. mutual dependency

True-False

1.	T		10.	T
2.	F		11.	F
3.	F		12.	F
4.	F		13.	F
5.	F		14.	F
6.	T		15.	T
7.	F		16.	T
8.	T		17.	T
9.	T		18.	T
			19.	F

Multiple choice

1.	b		14.	d
2.	a		15.	d
3.	b		16.	b
4.	d		17.	a
5.	c		18.	b
6.	a		19.	a
7.	a		20.	c
8.	d		21.	d
9.	d		22.	b
10.	d		23.	c
11.	c		24.	d
12.	b		25.	c
13.	c			

CHAPTER 4
Our Sexual Selves

Chapter Overview

This chapter examines the human sexual response as a potential for behavior that is shaped by a variety of factors. The chapter considers the various modes of sexual expression and the development of sexual preference. After pointing out the social and cultural forces that affect sexual attitudes and behaviors, the chapter considers issues such as sexuality throughout marriage, sexual dysfunctions, gay and lesbian sexualities, sexually transmitted disease, and sexual responsibility.

Chapter Summary

Social attitudes and values play an important role in the forms of sexual expression people find comfortable and enjoyable. Some of the changes of our times involve sexual expression, changing from the <u>patriarchal</u> <u>sexual</u> <u>script</u> toward the
_____ sexual script, with a variety of results: anxieties and satisfactions, opportunities and threats, experimentation and repression. Concepts such as _____, in which a husband forces his wife to engage in coitus with him against her will, fits less and less into the sexual script that seems to be increasingly dominant.

People express themselves sexually in many ways. Some of these ways (for instance, masturbation) have traditionally been stigmatized and others, such as cuddling and holding hands, have not been recognized as "sex."

Standards for sexual expression are changing, and the _____ <u>standard,</u> which specifies different sexual behavior for men and for women, is changing rapidly.

When sex is seen as a _____ <u>bond,</u> the partners commit themselves to expressing their sexual feelings to each other in an atmosphere of mutual cooperation. This may involve maintaining

a _____ view of sex, which sees sex as an
extension of the whole marital relationship rather
than as a purely physical exchange or only one
aspect of marriage. Observing these principles of
sexual expression may reduce sexual _____,
the sexual inadequacy inhibiting sexual desire,
response, excitement, or orgasm. Spectatoring may
occur less, and _____, spontaneous
sexual expression that is not focused on
performance, may happen more often.
     We know little about how sexual orientation,
the preference for a same-sex (_____)
or opposite-sex (_____) partner,
develops. It does not seem to result from family
relationships or other socialization.
     Conjugal sex changes throughout life. Young
spouses tend to place greater emphasis on sex than
do older mates. Developing habits of open
communication about sex helps a couple to meet
their changing sex needs as they get older. In
middle age, partners face changes in their
relationship. Wives often grow more assertive,
and sometimes this is accompanied by a changing
balance in the partners' relative sexual desire.
Many men and women remain sexually active into old
age.
     Sexual dysfunctions exist in many American
marriages. In general, treating the dysfunction
includes both therapeutic counseling to improve
the couple's relationship and sexual exercises
that the couple carries out in private.
     A factor to be considered when making
decisions about sexual encounters is sexually
transmitted disease. Though older forms of
sexually transmitted disease still exist and
continue as threats, newer ____ such as herpes and
_____ have intensified concern about STDs in
general. Some contemporary guidelines for dealing
with sexual relationships involve recognizing the
responsibilities of sexual relating and being open
and honest with oneself and with one's partner, to
facilitate _____ sex.
     The politics of STD encompasses the
possibilities of discrimination against certain
sexual, racial/ethnic, or other minorities. But
STD affects many people on more personal levels
than this.

Making sex a pleasure bond involves cooperation in a nurturing, caring relationship. To fully cooperate sexually, partners need to develop high self-esteem, to break free from restrictive gender-role stereotypes, and to communicate openly.

_____

Key Terms

You should be able to explain the concepts listed below. In your explanation, try to avoid using the concept you are explaining. You should be able to give several examples of each concept and to explain _why_ each example is an example.

patriarchal sexual script
   (p. 112)
expressive sexual script
   (p. 113)
marital rape
   (p. 113)
double standard
   (p. 115)
sexual responsibility
   (p. 120)
pleasure bond
   (p. 120)
holistic view of sex
   (p. 124)

sexual dysfunction
   (p. 130)
spectatoring
   (p. 130)
pleasuring
   (p. 130)
sexual orientation
   (p. 132)
heterosexual
   (p. 132)
homosexual
   (p. 132)
STDs
   (p. 136)
AIDS
   (p. 137)
safer sex
   (p. 141)

> **Point to Ponder:** You might make a list of the ten things that have had the greatest impact on personal relationships in the twentieth century--things that have affected the way people live with and relate to other people. Is AIDS on your list? It would be on many people's list. Think of the difference it has made in sex life. And that difference would be greater if more people changed their behavior in ways consistent with the AIDS problem. Of course, many have **not** changed their behavior.

Completion

Complete the following sentences by selecting the correct alternative from the terms listed above. Some may be used more than once. Some may not be used at all. Filling in a blank may require more than one word.

1. The concept of _____ refers to appropriate behaviors and attitudes for relationships between the sexes in which the man or men are viewed a more powerful than women, possessed of needs more pressing than those of women, and as being more central to relationships than are women.

2. The _____ refers to accepting one set of expected sexual behaviors for men and a different set of sexual behaviors for women.

3. Masters and Johnson define _____ as when an individual takes charge of his or her own sexual response.

4. The _____ insists that sex is not an isolated physical exchange, but is instead affected by the entire marriage relationship or, indeed, by all other aspects of a person's life.

5. If a person has some _____, then that person has some difficulty that inhibits sexual pleasure, excitement, and/or orgasm.

6. Critically viewing and evaluating one's own sexual activity--almost as if it were a product from an assembly line--in a way that may detract seriously from spontaneous enjoyment of sexual activity is called _____.

7. _____ are persons who are attracted to individuals of the opposite sex.

8. Persons who are sexually attracted to individuals of their own sex are _____.

9. AIDS and syphilis are examples of _____.

10. A serious sexual disease for which no medically approved cure is currently available is _____.

11. Uncertainties about the realities of AIDS-- and possibly of other STDs--has encouraged methods or techniques of _____.

_____

True-False

____ 1. About 70 - 80 percent of college women in the United States report having masturbated at some time in their lives.

____ 2. The matrilineal script is one of the two dominant "sexual scripts" in the United States.

____ 3. The traditional patriarchal script is one of the two dominant "sexual scripts" in the United States, and it is growing in power and ascendance against the other sexual script.

____ 4. Sexual adjustment is facilitated and made more efficient when spouses read from two opposite sexual scripts at the same time. It is impeded and made less efficient when the same sexual script is read by both partners.

____ 5. More married women are experiencing orgasm, and they are doing so more often.

____ 6. Sex therapists tend to see masturbation as unhelpful--but not seriously so--for healthy sexual development.

____ 7. Black men and black women appear no closer to each other in sexual attitudes than do white men and white women.

____ 8. Sexual responsibility means that one takes responsibility for the partner's achieving a satisfying sexual experience.

____ 9. Studies show that high self-esteem is conducive to having pleasurable sex.

____ 10. People with high self-esteem tend not to actively tell their sex partners what they find sexually pleasurable. They are secure enough not to need that kind of reassurance.

____ 11. The holistic view of sex is a perspective held by "macho" males who emphasize the patriarchal view of genital sex with females.

____ 12. Compared with the past, the percentage of homosexuals in our society today has probably not increased.

____ 13. A new wave of homophobia has resulted in increased hostility and violence toward gays.

____ 14. AIDS/HIV and other sexually transmitted diseases are medical conditions that have medical--but not social--consequences.

____ 15. People who get AIDS are often afraid of being the object of a variety of types of discrimination.

____ 16. Medication for AIDS care is expensive; the drug AZT, for example, costs about $6000 a year.

____ 17. There is at least one federally funded sex education course designed for junior high school students: Sex Respect.

____ 18. People should be honest with partners about their sexual history, AIDS status, and the like, but one's motives for engaging in sexual behavior are one's own and need not be shared.

Multiple Choice

1. The text specifically mentions which of the following as an example of self-disclosure?
   a. having a premarital physical examination
   b. telling one's partner the truth about one's previous sexual life
   c. having an orgasm with the partner
   d. having distant relatives find out about one's sex life

2. Sexual scripts:
   a. are plot-lines that couples innovatively write for their sexual activities
   b. come from the mass media and influence the manner in which people conduct their sexual lives
   c. suggest appropriate behaviors and attitudes regarding sexual behavior
   d. do not exist as guides for behavior, unlike social scripts, which let people know how to behave and feel

3. Which of the following **is** one of the sexual scripts discussed in the text?
   a. innovative sexuality
   b. controlled sexuality
   c. emancipated sexuality
   d. expressive sexuality

4. According to the text, sexually active teens:
   a. may be choosing by default
   b. have not proportionately increased during the last twenty years
   c. have declined by nearly one-half due to concerns about the AIDS epidemic
   d. are likeliest to come from sexually repressive parents

5. One study of people over age 60 found that _____ of them report engaging in masturbation.
   a. less than one-tenth
   b. nearly twenty percent
   c. nearly one-third
   d. just under one-half

6. Which of these is **not** one of the four standards of nonmarital sex?
   a. permissiveness with affection
   b. permissiveness without affection
   c. the double standard
   d. discrete sexual sharing

7. According to "the new double standard," both sexes:
   a. feel equally free to engage in sexual behavior, within reasonable guidelines
   b. feel free to engage in sexual behavior, and do so with an absence of guilt or shame, but feel that age differences about sexual expression are appropriate
   c. feel more permissive about their own sexual behavior but expect more conservative behavior from the opposite sex
   d. feel that active sexual expression is appropriate for both sexes, but only for those who have sex exclusively with one partner; inability to be sexually faithful is disapproved

8. Regarding sexual standards, research shows that _____ are more liberal than _____.
   a. whites; blacks
   b. the less educated; the more educated
   c. men; women
   d. North Americans; South Americans

9. Kerr, writing about asking for what we want in sex, offers some tips. One of them is to:
   a. ask for 100 percent of what you want in sex
   b. do not absolutely rule out the option of having more than one sexual partner
   c. realize that satisfying one's partner may require doing things that are personally distasteful, things that one would prefer not to do
   d. be reasonable and compromise to work out an equitable solution when there are disagreements

10. A study in the 1980s found that at least three times as many husbands as wives initiated sex, and the researchers found this pattern to be:
    a. the same for cohabitors as for married persons
    b. a decelerating trend
    c. clearer for whites and blacks than for Asian Americans
    d. true mainly for those in the middle-aged groups

11. Sex can be more effective as a(n) _____ when the relationship transcends traditional gender-role stereotypes.
    a. informal control mechanism
    b. tool to achieve micro-political aims
    c. reproductive option
    d. pleasure bond

12. Masters and Johnson speak of the "principle of mutuality" in sexual expression, and the essence of this principle is that the partners:
    a. share a common cause
    b. try to achieve the same levels of sexual satisfaction
    c. try to achieve simultaneous orgasm, though the levels of orgasmic experience may differ
    d. work equally hard at achieving satisfactory sexual expression

13. Eileen and David find their sexual relationship to be very satisfying. And the reason they find it that way is that they view their sexual relationship inseparable from the way their lives are entwined generally. Eileen and David are involved in:
    a. spectatoring
    b. interactive sexuality
    c. holistic view of sex
    d. none of these

14. Women's sexual desire tends to peak at about what age?
    a. in their late teens
    b. in the mid- to late 20s
    c. in their 30s
    d. during their mid-40s

15. Affected negatively or adversely by the double standard of aging are which of these?
    a. the unmarried
    b. the divorced
    c. older men
    d. women

16. The double standard of aging may produce the strongest general feelings of loneliness for which of the following categories?
    a. divorced but not remarried older men
    b. single older women
    c. married older women whose husbands are not in good health
    d. gay males older than age thirty

17. Eloise finds it difficult to relax and enjoy sexual experiences with her partner because she constantly worries about how well she is doing as a sex partner. It is as if she is trying to engage in sex while at the same time judging her sexual performance, and the judging interferes with her sexual performance. The concept that applies is:
   a. spectatoring
   b. refractive sexuality
   c. the "third person" effect
   d. the McCarey Syndrome

18. A(n) _____ is one who has a sexual preference for the opposite sex.
   a. heterosexual
   b. pedophile
   c. transsexual
   d. voyeur

19. Evidence indicates that _____ seem to broaden their sexual activity as they get older, i.e., engage more in sexual activity with their own sex, than do _____.
   a. women; men
   b. working-class persons; middle-class persons
   c. whites; blacks
   d. southerners; midwesterners

20. According to the text, every society contains about _____ who are homosexual.
   a. 1-3 percent
   b. 8-10 percent
   c. 12-18 percent
   d. 20-23 percent

21. The fact that every society has about the same percent of the population who are homosexual suggests that:
   a. sexual preference is the result of social learning
   b. a biological imperative is involved
   c. everyone has some tendency toward homosexuality, but it appears in only a minority
   d. heterosexuality is inherently the normal psychological pattern

22. A recent study of male cadavers, including homosexual males, found that one area of the body was significantly smaller among the gays than among the heterosexuals, namely:
    a. the breadth of the upper spinal column
    b. one or both testicles
    c. the femur
    d. an area in the brain

23. Studies of gays show that _____ are likely to have met through friendship networks rather than through more casual contacts.
    a. religiously involved gays
    b. suburban gays
    c. lesbians
    d. black gays

24. Which of the following stages of homosexual self-identity occurs during or after late adolescence?
    a. identity assumption
    b. identity confusion
    c. sensitization
    d. commitment

25. The four-stage model of homosexual self-identity may apply better to ___ than to ___.
    a. males; females
    b. whites; blacks
    c. the religious; the nonreligious
    d. younger adults; older adults

26. Women are now ___ of reported AIDS cases.
    a. 7 percent
    b. 13 percent
    c. 28 percent
    d. 37 percent

27. Of women who have AIDS, the largest percentage got it from:
    a. sexual contact with AIDS-infected men
    b. needle-sharing
    c. blood transfusions
    d. none of the above

28. Perhaps ____ of male homosexuals marry at least once, and in the process perhaps expose some heterosexual women to the AIDS virus.
    a. 10 percent
    b. 20 percent
    c. 35 percent
    d. 40 percent

29. HIV-positive employees are protected by ____ from being fired solely because of their HIV status.
    a. the Hill-Burton Act
    b. the Monmouth Provision of Title XXIII
    c. the Right-to-Work Law
    d. the Americans With Disabilities Act

30. In a recent study ____ of men reported lying to women about previous sexual relationships in order to have sex.
    a. 15 percent
    b. 35 percent
    c. 55 percent
    d. 65 percent

Your Opinion, Please: Someone once said that when sex is good, then it is 5 percent of a marriage. But if it is bad, then it is 95 percent of a marriage. What do you think such a viewpoint means? If you think it is true, **why** do you think it might be true? Try to give specific reasons.

Short-Answer Essay Questions

The following are sample short-answer essay questions--questions of the type you may be asked if your instructor uses questions like these. Even if your instructor does not use questions like these, you can help organize and consolidate your learning if you can answer these questions in a <u>well-organized</u> and <u>complete</u> manner. Do not think that brief essays are easier than lengthy essays. They can be more challenging because you are required to be brief yet complete. And after you have answered these, trying making up some similar questions and then answering them. Suggestion: The Study Questions at the end of each text chapter make very good short-answer essay questions.

1. Briefly explain the meaning of the concept "homosexualities" (note the plural).

2. Characterize or describe "sex in the middle years" as set forth in the text. In what ways is it different from sex earlier and later in life?

3. Make a list of the ways people can help protect themselves against getting the AIDS virus through sexual contact. For each item on the list, what are its drawbacks or insufficiencies?

‾‾‾‾‾
Essay

The following are sample essay questions-- questions of the type you may be asked if your instructor uses essay questions. Even if your instructor does not use essay questions, you can help organize and consolidate your learning if you can answer these questions in a well-organized and complete manner. Usually, the third essay question is the most challenging.

1. In what ways is sex related to self-esteem, pleasure, communication, and sharing?

2. Describe sexuality through the various stages of marriage. Be sure to use specific concepts and research results where appropriate to give additional credibility to your answer.

3. Is there a need for society--or its agents--to structure or in other ways shape the human sexual response? If yes, why? If no, why not?

Answers

Chapter Summary

expressive sexual script   **pleasuring**
such as **marital** **rape**   same sex **(homosexual)**
double standard   **(heterosexual)** partner
pleasure bond   **STDs** such as herpes
holistic view of sex   herpes and **AIDS**
reduce sexual **dysfunction**   facilitate **safe(r)** sex

Completion

1. patriarchal sexual   6. spectatorship
      script   7. heterosexuals
2. double standard   8. homosexuals
3. sexual responsibility   9. STDs
4. holistic view of sex   10. AIDS
5. sexual dysfunction   11. safer sex

True-False

1. T   10. F
2. F   11. F
3. F   12. T
4. F   13. T
5. T   14. F
6. F   15. T
7. T   16. F
8. F   17. T
9. T   18. F

Multiple Choice

1.	c	16.	b
2.	c	17.	a
3.	d	18.	a
4.	a	19.	a
5.	d	20.	b
6.	d	21.	b
7.	c	22.	d
8.	c	23.	c
9.	a	24.	a
10.	a	25.	b
11.	d	26.	b
12.	a	27.	b
13.	c	28.	b
14.	c	29.	d
15.	d	30.	b

CHAPTER 5
Being Single: Alone and with Others

_____

Chapter Overview

Being single is more frequent in American society.
The never-married, the widowed, and the divorced
have become a more common part of life in our
society. Many people postpone marriage until they
are older.   Some never marry, choosing lifelong
singlehood instead.   Some would prefer to marry
but remain single nevertheless. Attitudes toward
singlehood are changing, making this choice a more
legitimate one rather than deviant. Some remain
with parents, some live in group or communal
arrangements, and others cohabit.   These changes
have a variety of consequences and significant
implications for individuals, families, life-
styles, organizations, and society itself.

_____

Chapter Summary

Since the 1960s the number of people chosing
to live alone--or, _____ --has risen.
Much of this increase is due simply to rising
numbers of young adults who are typically single.
Although there is a growing tendency for these
young adults to postpone marriage until they are
older, this is not a new trend but rather a return
to a pattern typical early this century.
    One reason people are postponing marriage
today is that increased job opportunities for
women may make marriage less attractive to them.
Also, the low sex ratio has caused a number of
women to postpone marriage or put it off entirely.
And attitudes toward marriage and singlehood are
changing so that now being single is viewed not so
much as deviant but as a legitimate choice.
    Singles can be classified according to
whether they freely choose this option for a short
time (_____singles), plan to
remain single because they want to (voluntary
_____ singles), would prefer to be
married but currently are not (_____
_____ singles), or remain single even

though they would prefer to be married
(involuntary _____ singles).
    Singles used to live primarily in cities,
often because employment and leisure opportunities
drew them there, but recently more singles have
moved to the suburbs. More and more singles are
living in their parents' homes; this is usually at
least partly a result of economic constraints.
Some singles have chosen to live in group homes or
in a _____.   A substantial number of
unmarrieds are cohabiting, or "living together."
Of these, some are heterosexual couples and some
are gay or lesbian. And some women share a man in
a polygamous arrangement.
    Although research finds married persons
physically and psychologically healthier and
happier than singles, this may change as
singlehood allows the sexual expression once
reserved for marriage and as marriage fails to
guarantee the security it once did.

_____
Key Terms

You should be able to explain the concepts listed
below. In your explanation, try to avoid using
the concept you are explaining. You should be
able to give several examples of each concept and
to explain why each example is an example.

single
   (p. 152)
voluntary temporary
   singles
   (p. 163)
voluntary stable
   singles
   (p. 164)

involuntary temporary
   singles
   (p. 164)
involuntary stable
   singles
   (p. 164)
commune
   (p. 168)

```
Point to Ponder:   If someone important to you
                   is sick or dies, do you
think that you ought to be able to take time
off from work to attend to personal concerns?
This time off might or might not be paid.
In either case, most people have no problem
arranging for some time off when their
husbands, wives, sons, or daughters are sick
or in need of help.  But many singles sometimes
find it difficult to get employer approval
for time off from work to help out a roommate
or person sharing their apartment when the
need arises.  The problem, apparently, is
their marital status of singlehood.  Policy
in many instances does not match personal
realities in our society.  Should policies be
changed so that singlehood does not result in
discrimination--often discrimination with
serious economic consequences.   Why, specifi-
cally, do you answer this question as you do?
```

Completion

Complete the following sentences by selecting the
correct alternative from the terms listed above.
Some may be used more than once. Some may not be
used at all. Filling in a blank may require more
than one word.

1. _____ includes the following
    categories: the divorced, separated, widowed,
    or never-married.

2. Living in a _____ involves sharing a
    group household with others unrelated by
    blood, marriage, or adoption.

3. _____
    are younger never-marrieds and divorced
    persons who are postponing marriage or
    remarriage. For them, searching for a mate
    has a lower priority than other activities
    such as a career.

4. _____
   include cohabitants who do not intend to marry
   and those whose life-styles preclude marriage,
   such as priests and nuns.

5. _____
   are singles who would like--and expect--to
   marry and are actively seeking mates.

6. _____
   are those persons who wanted to marry, never
   did, and have now come to accept being single
   as their probable life situation.

7. Kenneth decided to devote his life to
   maintaining and increasing his investments in
   the stock market. He felt he didn't have time
   for any continuing relationships. He never
   wanted to marry and never did, devoting his
   time instead to economic advancement. Kenneth
   is an example of a(n) _____
   _____.

8. Frank graduated from high school, met and went
   out with Marie while they both attended
   college, and both agreed that they could spend
   their lives together, as both had always
   wanted marriage and children. For now,
   though, they feel it is necessary to postpone
   marriage until both finish college. Frank and
   Marie are examples of _____
   _____.

9. Karen would like to marry, but has not been
   able to find the kind of man with whom she
   feels she could be happy. She continues to
   try, however, and hopes to marry someday. For
   now, she lives alone. Karen is an example of
   a(n) _____.

10. Joel spent his early and middle years trying
    to invent a way to separate aluminum from
    aluminium oxide. Now, in his later years, he
    finds himself unable to realize his dream of
    an eventual marriage. He is coming to accept
    the fact that he will never marry and is in
    the process of--reluctantly--settling into
    this situation. Joel is an example of a(n)
    _____.

You should be familiar with the main question being investigated and the research results for the following studies:

    Marital status of the U.S. population
    Cargan and Melko: Myths and realities about
       singles
    Staples: research on singlehood among blacks
    Scott: polygamy among American blacks

True-False

____ 1. Since 1960, singlehood has been on the increase in the United States.

____ 2. In 1990, about 79 percent of males age 20 to 24 were single.

____ 3. Recently there has been a trend toward getting married at a later age as compared to the 1950s.

____ 4. A majority of singles plan to marry.

____ 5. The proportion of blacks who are married has been declining.

____ 6. There are more black women than black men in the USA, partly because of higher mortality rates for black males.

____ 7. Changing attitudes toward marriage and singlehood are among the reasons for the increase in singlehood.

____ 8. According to the text, singles are a highly varied group.

____ 9. Whites spend more time in a marriage than do blacks, but blacks spend more time as singles.

_____ 10. Alice hopes to marry as soon as she can, but hasn't found the right person. She continues to hope, however, that someday she will marry. Alice is an example of a voluntary stable single.

_____ 11. Alexander decided to devote his life to developing the next generation of computer programming language. He consciously decided to stay single forever so he could devote himself to this task. Alexander is an example of a voluntary stable single.

_____ 12. Staples developed a typology in which he described six types of singles.

_____ 13. Married people are likelier to have white-collar jobs and higher incomes than are singles, regardless of age and education.

_____ 14. Research in the 1980s pointed to job discrimination against singles, especially against single men.

_____ 15. Single women buyers represent the fastest growing segment of the housing market.

_____ 16. In the recent past, more and more singles are living with one or both parents.

_____ 17. Single adult males are likelier to be living in the parental home than are single adult females.

_____ 18. The text states that communes are among the most successful enemies of singlehood, as seen by the kibbutz and by the Oneida colony.

_____ 19. The majority of cohabiting relationships are relatively short term.

_____ 20. African-Americans have cohabitation rates three times greater than whites.

_____ 21. According to the text, many state or local laws still forbid sex between persons not married to each other.

_____ 22. Unmarried couples generally find it difficult to open joint charge accounts.

_____ 23. So far no courts have permitted a lesbian partner to adopt the biological child of the other partner, to ensure legal parenthood to both members of the couple raising the child.

_____ 24. There is no such legal status as "domestic partner."

_____ 25. According to the text, in other than legal respects, gay and lesbian relationships are similar to heterosexual ones.

Multiple Choice

1. Which of these is NOT included in the category termed "single"?
   a. divorced
   b. separated
   c. widowed
   d. none: all are included

2. There are now approximately _____ single persons in the United States.
   a. 10 million
   b. 34 million
   c. 68 million
   d. 110 million

3. According to the U.S. Bureau of the Census, singles now constitute about _____ percent of the general population.
   a. 15
   b. 25
   c. 38
   d. 55

4. _____ are much more likely to be widowed than are _____.
    a. The middle class; the working class
    b. Methodists; Baptists
    c. Women; men
    d. The well educated; the less educated

5. Staples's research reports that blacks:
    a. place a high value on marriage
    b. place a low value on marriage
    c. spend more years in the "married" status than do whites
    d. have a lower rate of desertion than do Hispanics

6. David likes being married. He was saddened and disorganized when his wife died. He wants to be married again. This situation illustrates which of these?
    a. voluntary temporary single
    b. voluntary stable single
    c. involuntary temporary single
    d. involuntary stable single

7. The proportion of married blacks in the general population has:
    a. declined sharply
    b. increased sharply
    c. increased slightly
    d. declined slightly

8. The text states that singles may feel _____ about being single or being married.
    a. despondent
    b. deeply fulfilled
    c. persistent annoyance
    d. ambivalent

9. The evidence indicates that singles tend to move:
    a. to better and better apartments
    b. to more populated residence areas
    c. away from areas of high population turnover
    d. away from areas of high crime rates

10. The U.S. Census estimates that just over ___ of all couples are living together, or cohabiting.
    a. 76 percent
    b. 53 percent
    c. 36 percent
    d.   5 percent

11. According to the text, about _____ of cohabiting couples eventually marry.
    a. one-half
    b. one-fourth
    c. one-sixth
    d. one-third

12. Unmarried couples are more likely than married couples to be:
    a. older
    b. previously married
    c. interracial
    d. widowed

13. Singles reportedly have _____ than have married persons.
    a. more feelings of loneliness
    b. higher rates of religious participation
    c. less feelings of alienation
    d. lower accident rates

14. The marriage gradient refers to which of the following?
    a. people who are single because they "can't make the grade" in the dating process
    b. people marrying from within their own age group
    c. females marrying men who live up to their standards of excellence in potential mates
    d. men marrying women from a social class lower than their own

15. Which of these illustrates a temporary voluntary single?
    a. Patrick, who is single and intends to remain that way until he is "trapped" by some woman's pregnancy and threat of a paternity suit
    b. Carolyn, who is a professional woman waiting for "the right man" to marry
    c. Jack, who wants a wife but cannot find anyone who will marry him
    d. Margaret, whose husband deserted her because she wouldn't change her ways to suit him

16. The largest category of young adults living with their parents is:
    a. children whose parents cannot afford to have the children with them
    b. never-married young adults who for various reasons continue to live with their parent(s)
    c. children whose parents do not want to have their children living with them
    d. young couples who live with a parent because of the parent's poor health

17. More ___ than ___ live with their parents.
    a. religious persons; nonreligious persons
    b. southerners; midwesterners
    c. men; women
    d. divorced; separated

18. The number of unmarried couples living together in the United States has:
    a. increased dramatically since 1960
    b. not increased since the mid-1970s
    c. actually declined somewhat since 1985
    d. been generally stable since 1980

19. According to the text, the largest percentage of cohabitants in the United States, in 1989, was in which of the following age groups?
    a. under age 25
    b. age 25-44
    c. age 35-44
    d. age 45-64

20. According to the text, cohabitants are likely to be all of the following EXCEPT which one?
    a. less actively religious
    b. more politically liberal
    c. living in large urban areas
    d. living in the Midwest and South

21. Unmarried couples are more likely than married couples to be:
    a. older
    b. third or fourth partner relationships
    c. interracial
    d. deeply conflict-habituated

22. Almost _____ the states in the United States have laws against "sodomy" or "deviant sexual intercourse," forbidding homosexual relationships.
    a. one-sixth
    b. one-fourth
    c. one-half
    d. three-fourths

23. Regarding taxes, it is _____ for unmarried couples to file a joint income-tax return.
    a. increasingly common
    b. more infrequent
    c. tax-disadvantageous
    d. illegal

24. According to the text, a status that is less and less common in the United States is:
    a. common law marriage
    b. legal separation
    c. domestic partnership
    d. cohabitation

25. Meeting the needs of many gay or lesbian couples--and many heterosexual couples as well--is:
    a. common law marriage
    b. domestic partnership
    c. nominal affiliation agreements
    d. ex post facto marriage

26. After comparing gay men, heterosexual men, lesbians, and heterosexual women, Peplau concluded there was little evidence for a distinctive homosexual "ethos" or orientation toward:
    a. desirability of children
    b. gender-based division of household labor
    c. agreement with basic American cultural norms
    d. love relationships

27. There is evidence that never-married females' overall satisfaction:
    a. declines after about age 30
    b. increases after about age 45
    c. declines after about age 45
    d. is as stable as that of married females

28. The text states that _____ is a factor that affects satisfaction with singlehood.
    a. physical well-being
    b. sexual orientation
    c. physical appearance
    d. choice

29. The "new other woman" refers to:
    a. a woman who remains steadfastly single
    b. single women's sexual relationships with married men
    c. the woman loved by two men who share her
    d. bisexuality, or shared lesbian relationships

30. The text states that _____ is important to many older single women's sense of identity with groups larger than their own families.
    a. organized leisure
    b. shopping and buying
    c. volunteerism
    d. formal education

> **Your Opinion, Please:** Some governments give
> tax incentives to
> married people to encourage formation of
> married households. Do you think that the
> perceived needs of government justify this
> kind of policy? Specifically, why?

Short-Answer Essay Questions

The following are sample short-answer essay
questions--questions of the type you may be asked
if your instructor uses questions like these.
Even if your instructor does not use questions
like these, you can help organize and consolidate
your learning if you can answer these questions in
a <u>well-organized</u> and <u>complete</u> manner. Do not
think that brief essays are easier than lengthy
essays. They may be more challenging because you
are required to be brief yet complete. And after
you have answered these, try making up some
similar questions and then answering them.
Suggestion: The Study Questions at the end of each
text chapter make very good short-answer essay
questions.

1. List at least five of the seven myths about
    singles discussed in the text.

2. In what ways are black singles different from
    white singles?

3. Why do many young singles in the United States choose to continue to live with their parents?

4. What special financial problems face persons who are legally unrelated but who live together?

Essay

The following are sample essay questions--questions of the type you may be asked if your instructor uses essay questions. Even if your instructor does not use essay questions, you can help organize and consolidate your learning if you can answer these questions in a <u>well-organized</u> and <u>complete</u> manner. Usually, the third essay question is the most challenging.

1. Summarize with supportive detail the income and residential patterns of singles.

2. What are the positive and negative aspects of singlehood, according to the text?

3. In what ways are the life satisfactions and problems of single men different from those of single women? Use research results and data to give your answer greater credibility.

Answers

Chapter Summary

or, **singles** --has risen
voluntary temporary singles
voluntary **stable** singles
involuntary temporary singles
involuntary **stable** singles
or in a **commune**

Completion

1. singles
2. commune
3. voluntary temporary
   singles
4. voluntary stable
   singles
5. involuntary temporary
   singles
6. involuntary stable
   singles
7. voluntary stable
   single
8. voluntary temporary
   singles
9. involuntary
   temporary single
10. involuntary stable
    single

True-False

1.	T	9.	T	17.	T
2.	T	10.	F	18.	F
3.	T	11.	T	19.	F
4.	T	12.	F	20.	T
5.	T	13.	T	21.	T
6.	T	14.	T	22.	T
7.	T	15.	T	23.	F
8.	T	16.	T	24.	F
				25.	T

Multiple Choice

1.	d	11.	a	21.	c
2.	c	12.	c	22.	c
3.	c	13.	a	23.	d
4.	c	14.	d	24.	a
5.	a	15.	b	25.	b
6.	c	16.	b	26.	d
7.	a	17.	c	27.	a
8.	d	18.	a	28.	d
9.	b	19.	b	29.	b
10.	d	20.	d	30.	c

CHAPTER 6
Choosing Each Other

Chapter Overview

This chapter examines the social forces that influence the choices people make--knowledgeably or by default--in selecting people to date, cohabit with, and marry. After pointing out that we tend to form relationships with those who are in many ways similar to ourselves, the chapter examines the roles played by physical attractiveness and rapport in encouraging various forms of courtship, cohabitation, and marriage. The chapter ends by examining the factors that make for stability or instability in marriages and explores the variety of reasons people have for marriage.

Chapter Summary

Many--perhaps most--Americans think of marriage as going with love like a "horse and carriage," but this association is virtually unique to our modern culture. Historically, marriages were often arranged in the marriage market, as business deals. Many elements of the basic exchange (a man's providing financial support in exchange for the woman's child-rearing capabilities, domestic services, and sexual availability) remain, sometimes complementing and sometimes conflicting with other parts of our culture and society.

What attracts people to each other? Three important factors are _____ or pressures to marry outside one's own social group, _____ or pressures to marry within one's social group, and physical attractiveness. Some elements of homogamy are _____, social pressure, feeling at home with each other, and the fair exchange. Three patterns of courtship familiar in our society are _____, getting together, and _____.

Besides homogamy and the degree of intimacy developed during courtship, two other factors related to the success of a marriage are a couple's age at marriage and their reasons for marrying. People who marry too young or too old are less likely to stay married; and there are several negative reasons for marrying that can lead to unhappiness or divorce.

If potential marriage unhappiness can be anticipated, breaking up before marriage is by far the best course of action, however difficult it seems at the time. A certain number of courting relationships will end in this fashion.

But many couples will go on to marry. With this chapter, the text begins to examine which form the marriage is likely to take and some choices couples make in setting up marriages.

Key Terms

You should be able to explain the concepts listed below. In your explanation, try to avoid using the concept you are explaining. You should be able to give several examples of each concept and to explain why each example is an example.

courtly love
  (p. 193)
dowry
  (p. 193)
marriage market
  (p. 193)
exchange theory
  (p. 193)
sex ratio
  (p. 197)
pool of eligibles
  (p. 198)
homogamy
  (p. 198)
endogamy
  (p. 198)
heterogamy
  (p. 198)

marriage gradient
  (p. 198)
hypergamy
  (p. 201)
courtship
  (p. 207)
imaging
  (p. 209)
dating
  (p. 210)
cohabitation
  (p. 214)
getting together
  (p. 214)
two-stage marriage
  (p. 214)
date rape;
    acquaintance rape
    (p. 212)

Point to Ponder: In the United States, some states allow a person to marry a first cousin, while other states prohibit such marriage. Similarly, some states do not allow adopted brothers and sisters to marry. What do these states accomplish by these restrictions? Why do you come to the conclusions you do?

Completion

Complete the following sentences by selecting the correct alternative from the terms listed above. Some may be used more than once. Some may not be used at all. Filling in a blank may require more than one word.

1. Relationships based on _____ required a great deal of idealization and were not necessarily sexually consummated.

2. A _____ is a sum of money or property brought to the marriage by the female.

3. Viewing marriage in terms of bargaining, a marketplace, and resources is a central part of what is called the _____.

4. The _____ is the number of men to women in a society or subgroup of society.

5. The idea that people who date or are going together come to the relationship with their assets and liabilities and make the best deal they can is part of what is meant by the _____ perspective.

6. Those persons who have the personal, social, and other characteristics to make them an appropriate marriage choice can be termed a _____.

7. _____ exists when people marry others of similar race, age, education, religious background, and social class.

8. _____ occurs when people marry within their own social group.

9. _____ refers to marriage to one who belongs to a higher socioeconomic class than one's own.

10. _____ occurs when people marry outside their social group.

11. _____ refers to marriage between those who are different in race, age, education, religious background, or social class.

12. _____ can be a courtship process in which groups of women and men congregate at a party or share an activity, but there is relatively little pressure to relate to any one member of the opposite sex.

13. When unrelated persons of the opposite sex live together without being married, the relationship is termed _____.

Key Research Studies

You should be familiar with the main question being investigated and the research results for the following studies:

    Ridley: motives and experiences of cohabitors
    Libby: the "getting together" process
    Macklin: cohabiting couples
    National Center for Health Statistics:
      cohabitation and marital histories

Key Theories

Exchange theory of dating and marriage
Theories of homogamy, heterogramy; endogamy; exogamy; hypergamy

____ 1. Courtly love (or romantic love) involved a great deal of idealization and was not necessarily sexually consummated.

____ 2. A "dowry" is an older woman, usually unmarried, who has considerable wealth.

____ 3. Exchange theory of mate selection assumes that people are largely rational.

____ 4. The sex ratio refers to the ratio of successful sexual encounters to unsuccessful sexual encounters.

____ 5. The pool of eligibles refers to the number of now-available divorced men in a population.

____ 6. Propinquity refers to one's level of susceptibility to romantic or erotic stimuli.

____ 7. Heterogamy refers to marriage between persons of the opposite sex.

____ 8. When people are trying to appear to be "unrealistically attractive and nice" and the date is going likewise--this is what some sociologists call "imaging."

____ 9. Hypergamy refers to a person marrying someone because of deep anxiety that he or she is about to end up being single for the rest of his or her life.

____ 10. Research indicates that heterogamous marriages are more stable than those that are not heterogamous.

____ 11. The second stage of Mead's "two-stage marriage" is what she called "parental marriage."

___ 12. A nationwide survey (1988) found that cohabitation was more likely to progress to marriage among black women than among white women.

___ 13. About three-fifths of cohabitors eventually marry each other.

___ 14. Researchers have found a significant difference in the marriages of those who cohabited before marriage and those who did not.

___ 15. Samuel, a black man, married a black woman of his own social class and religion. This is an example of what sociologists call "exogamy."

___ 16. Louise, a young white female from a working-class background, married David, an older black man who is a financially well-off attorney. This is an example of hypergamy.

Multiple Choice

1. "Romantic love" is closest in meaning to which of these?
   a. poetic bargaining
   b. the heraldic code
   c. courtly love
   d. Napoleonic love

2. The exchange theory of dating and mate selection is closest in meaning and process to which of these?
   a. the sex ratio
   b. the marriage market
   c. gender roles
   d. imaging

3. The sex ratio is calculated in terms of:
   a. persons' levels of physical beauty
   b. a kind of "batting average" regarding sexual conquests
   c. the extent to which men and women embody ideals of masculinity and femininity
   d. the number of males per 100 females

4. Which of these means "marrying outside one's group"?
   a. homogamy
   b. inappropriate pool of eligibles
   c. exogamy
   d. hypergamy

5. A black Roman Catholic marries a white Buddhist. This is an example of:
   a. heterogamy
   b. hypergamy
   c. endogamy
   d. propinquity

6. Andy is a black banker who marries Sara, a white bank teller. Sara feels she has gone up in social class, while Andy feels he has gained entrance to the white world. This illustrates:
   a. hypergamy
   b. homogamy
   c. endogamy
   d. propinquity

7. "Getting together" and "dating" are different from each other in that dating:
   a. has more stress
   b. is less reliable in terms of having a fun time
   c. is something that "just happens" whereas getting together is more planned
   d. automatically has a weaker connotation of leading to a more serious relationship

8. "Getting together" is viewed by the text as:
   a. most typical of the working class
   b. producing less favorable outcomes than dating produces
   c. an alternative script to dating
   d. more productive of marriages than is traditional dating

9. Mead's two-stage marriage consists of _____ marriage and parental marriage.
   a. affectionate
   b. quasi
   c. individual
   d. nontraditional

10. The text agrees with the wheel theory of love when it states that _____ is a necessary first step to beginning most relationships.
    a. flexibility
    b. rapport
    c. liberality
    d. mutual trust

11. A comparison of the difference in the marriages of those who cohabited before marrying and those who did not showed:
    a. the cohabitors were much more adjusted than were the noncohabitors
    b. the cohabitors were moderately more adjusted than were the noncohabitors
    c. very little difference between the two groups
    d. the noncohabitors were moderately more adjusted than were the cohabitors

12. When researchers asked students their reasons for cohabiting, they found that men's and women's goals differed. Men were more likely to indicate _____ as their reason.
    a. sexual gratification
    b. having someone to help with work around the house or apartment
    c. companionship
    d. trial marriage

13. According to one study, breakups were shown to occur more often:
    a. when the partners were together for uninterrupted long periods of time
    b. between partners who were not equally involved in the relationships
    c. when relationships were homogamous
    d. between partners each of whom had some solitary leisure pursuits

14. According to a study of college students in the Boston area:
    a. few breakups were mutual
    b. most partners saw a breakup coming at least a year before the breakup happened
    c. most partners were surprised or startled that the breakup happened
    d. most breakups happened during the middle of the term, quarter, or semester

15. Which of the following is **not** one of the four common patterns of cohabitation detected by Carl Ridley and his colleagues?
    a. "Linus blanket"
    b. emancipation
    c. power gratification
    d. convenience

16. The text suggests that he or she may not be "right for you" if he or she does not:
    a. initiate some type of contact at least once daily
    b. have other friends
    c. have the same sleeping/waking hours as you have
    d. always make you feel like having erotic or sexual relations when with him/her

17. The text suggests that, regarding taboo topics that come up for discussion with a person who is "right for you":
    a. it is best to discuss such topics when a mediator is present
    b. things tend to go more smoothly when such topics are avoided or ignored
    c. taboo topics are strictly private matters and can be discussed with persons other than the partner if discussion is needed
    d. topics that have a bearing on the relationship are never taboo

18. In 1990 the median or "average" age at first marriage was about _____ for men and _____ for women.
    a. 20.2; 18.5
    b. 23.5; 21
    c. 24.8; 22
    d. 26.1; 23.9

19. According to the text, the most significant distinction between stable and unstable marriages is the distinction between _____ and other marriages.
    a. teenage
    b. educationally homogamous
    c. economically homogamous
    d. religious partners

20. Which of the following is associated with being married at an earlier age?
    a. partners' religious involvement
    b. having parents who are divorced
    c. having one or no siblings
    d. low socioeconomic origins

21. The text cites four elements of maturity that sociologist David Knox considers to be critical for marital stability. Which of the following is NOT one of those four?
    a. emotional maturity
    b. relationship maturity
    c. economic maturity
    d. societal maturity

22. The text suggest that which of the following may be a variable predicting or determining a relationship's potential for success?
    a. marriage gradient scores
    b. age
    c. extent to which ethnicity/race is important to the partners
    d. stability of social class

23. Research indicates that for both black and white couples, _____ is most associated with subsequent marital breakup.
    a. disagreements about degree of religious involvement
    b. age differences
    c. childbirth before marriage
    d. parental disapproval of the relationship

24. According to the text, for teenage parents, babies:
    a. help cement an otherwise shaky relationship
    b. can interfere with achieving educational and career goals
    c. usually serve as a test of whether or not the couple have a relationship worth keeping
    d. often keep the grandparents in contact with each other if/when divorce occurs

25. According to the text, physical appearance, social pressure, and economic advancement as reasons for marriage are:
    a. some things that should be considered before legal union is seriously contemplated
    b. associated with unstable marriage relationships
    c. topics that should be included in any reputable premarital counseling session
    d. openly considered by lower-class partners but only subconsciously considered by partners in the middle-class and above

26. According to the text, sociologist David Knox lists three positive bases for a happy marriage. Which of these is NOT one of those three reasons?
    a. companionship
    b. emotional security
    c. desire to parent and raise children
    d. social pressure

Your Opinion, Please: Some research shows that cohabitation--or, living together before marriage--does not improve a couple's level of adjustment once they are married. It is an interesting finding because many cohabitors consider cohabiting a "trial marriage." Can you think of any reasonable reasons why these research results turn out as they do? Try to think of **specific** reasons.

Short-Answer Essay Questions

The following are sample short-answer essay questions--questions of the type you may be asked if your instructor uses questions like these. Even if your instructor does not use questions like these, you can help organize and consolidate your learning if you can answer these questions in

a <u>well-organized</u> and <u>complete</u> manner. Do not
think that brief essays are easier than lengthy
essays. They can be more challenging because you
are required to be brief yet complete. And after
you have answered these, try making up some
similar questions and answering them. Suggestion:
The Study Questions at the end of each text
chapter make very good short-answer essay
questions.

1. Differentiate between endogamy and homogamy.
    Give an example of each and make it clear why
    your example is a good example.

2. How is <u>cohabitation</u> different from a <u>two-stage</u>
    <u>marriage?</u>

3. Distinguish between <u>dating</u> and <u>getting</u>
    <u>together,</u> as explained in the text.

Essay

The following are sample essay questions--
questions of the type you may be asked if your
instructor uses essay questions. Even if your

instructor does not use essay questions, you can help organize and consolidate your learning if you can answer these questions in a <u>well-organized</u> and <u>complete</u> manner. Usually, the third essay question is the most challenging.

1. Explore the extent to which mate selection is a "bargaining" relationship. Support your answer with pertinent data and research findings.

2. Explain reasons why "Jeff" married someone who is white (just like him), is Roman Catholic (just like him), is about age 23 (just like him), is at the same college he attends, and takes classes in the same academic building in which he takes classes.

3. Your friend confides in you that he/she seems to be in the process of cohabiting and knows nothing about this way of living, but wants to know "the facts" about the effects of cohabitation on a relationship. Summarize the facts about cohabitation in a way that would make sense to your friend. Where possible, make specific references to research results or findings.

Answers

Chapter Summary

<u>exogamy,</u> or pressures to marry outside
<u>endogamy,</u> or pressures to marry within
some elements of homogamy are <u>propinquity</u>
familiar in our society are <u>dating</u>
getting together, and <u>cohabitation</u>

Completion

1.	courtly love	7.	homogamy
2.	dowry	8.	endogamy
3.	marriage market	9.	hypergamy
4.	sex ratio	10.	exogamy
5.	exchange theory	11.	heterogamy
6.	pool of eligibles	12.	getting together
		13.	cohabitation

True-False

1.	T	9.	F
2.	F	10.	F
3.	T	11.	T
4.	F	12.	F
5.	F	13.	F
6.	F	14.	F
7.	F	15.	F
8.	T	16.	T

Multiple Choice

1.	c	14.	a
2.	b	15.	c
3.	d	16.	b
4.	c	17.	d
5.	a	18.	d
6.	a	19.	a
7.	a	20.	d
8.	c	21.	d
9.	c	22.	b
10.	b	23.	c
11.	c	24.	b
12.	a	25.	b
13.	b	26.	d

CHAPTER 7
Marriage: A Unique Relationship

Chapter Overview

This chapter examines the various types of
relationships existing within marriages.
Marriage relationships need not be of only one
type. The institution seems to be capable of
meeting human emotional, social, and other needs
by being flexible. And marriages do change.
Also, the early stages of a marriage are typically
different from the middle and later stages of
marriage. This chapter explores the interior
structure of various types of marriages found in
our society.

Chapter Summary

Although marriage is less permanent and more
flexible than it has ever been--and perhaps
couples today are more in need of effective
premarriage counseling--marriage is still set
apart from other human relationships. The
marriage _____ includes expectations of
permanence and primariness. As both of these
expectations come to depend less on legal
definitions and social expectations, partners need
to invest more effort in sustaining a marriage.

Two opposite poles on a continuum of marriage
are the <u>utilitarian</u> and the _____ marriage.
Most real marriages fall somewhere in between.
Some frequently occurring marital types are the
<u>conflict-habituated,</u> the _____, the
<u>passive-congenial,</u> the _____, and the
_____ marriage.

In light of the many choices and styles of
marriage observed through the life cycle, some
people are asking serious questions about many
issues. Partners change over the course of a
marriage, so a relationship needs to be flexible
if it is to continue to be intrinsically
satisfying. <u>Static</u> marriages are usually
devitalized. <u>Marriage</u> or _____, which can
usually be negotiated as the need arises, are one
useful way of coming to mutual agreement. Working

on marriage agreements together can help partners develop a "couple identity"--one of the tasks of early marriage.

Key Terms

You should be able to explain the concepts listed below. In your explanation, try to avoid using the concept you are explaining. You should be able to give several examples of each concept and to explain <u>why</u> each example is an example.

marriage premise
    (p. 236)
primariness
    (p. 237)
sexual exclusivity
    (p. 238)
intrinsic marriage
    (p. 245)
utilitarian marriage
    (p. 245)
conflict-habituated
  marriage
    (p. 246)
devitalized marriage
    (p. 246)

passive-congenial
  marriage
    (p. 247)
vital marriage
    (p. 247)
total marriage
    (p. 248)
static marriage
    (p. 250)
flexible marriage
    (p. 250)
personal marriage
  agreement
    (p. 251)
relationship agreement
    (p. 251)

Point to Ponder: In this chapter, you will find out about the many ways in which couples live their lives. Some couples revel in each other's lives and fulfill each other. Other couples seem to be friends, but little else. And there are other types of couples as well. Question: What **should** marriage "be"? Should marriage be a deeply fulfilling experience, or is a pleasant friendship enough? Put differently, what is "adjustment" in marriage? Is a spouse well adjusted if she/he is happy and fulfilled, but the partner is miserable? For some couples, should an uneasy truce be "as good as it gets"?

Completion

Complete the following sentences by selecting the correct alternative from the terms listed above. Some may be used more than once. Some may not be used at all. Filling a blank may require more than one word.

1. A _____ is one begun or maintained for primarily practical purposes.

2. Partners in a _____ allow and encourage the partner to grow and change, with roles being negotiated as the needs of each partner changes.

3. _____ requires the commitment of both partners to keep each other the most important persons in their lives.

4. Primariness is strongly connected with the idea of _____.

5. Permanence and primariness are the two important elements of _____.

6. Henriette's job demands a lot of her time, and she has to spend much of this time with her colleague, Joe. But her husband Ted isn't jealous because he realizes that he is absolutely #1 in Henriette's mind. And it is for him--and her--long-term welfare that she spends so much time at work. This "being #1 in her mind" is what the text means when it uses the term _____.

7. In _____ partners acknowledge their incompatibility and recognize the atmosphere of tension as normal.

8. Partners in _____ rely on their formal, legal bond to enforce permanence, strict monogamy, and rigid husband-wife role behavior.

9. Usually couples agree that _____ will include expectations of sexual exclusivity, in which partners promise to have sexual relations only with each other.

10. David and Paula have been married for ten years. David is very happy with the relationship. Paula is there for him as a homemaker, sexually, and as a companion for social gatherings. She occupies his otherwise available time with pleasant conversation and interesting activities. She meets his daily needs. He likes to think that he meets hers. Theirs is an example of a(n) _____ marriage.

11. Partners in _____ marriages never expected marriage to encompass emotional intensity. Instead, they stress the "sensibility" of their decision to marry.

12. Unlike utilitarian marriages, _____ marriages are held together primarily by mutual affection and intimacy. People's high expectations about marriages of this type are often unrealistic and, despite partners' closeness and empathy, there can be no guarantee that the intimacy will continue indefinitely.

13. _____ is a type of marriage that is almost all-encompassing, with many points of vital meshing to the extent of leaving few aspects of personal life unshared.

14. Partners in _____ feel that the relationship has lost its original zest, intimacy, and meaning. Cuber and Harroff found this type of marriage to be exceedingly common among their respondents.

15. In _____, there is communication, negotiating, and coming to an agreement on behavioral expectations in the relationship.

16. In _____, the
    partners intensely enjoy being together.
    There are few areas of tension in such
    marriages, but the partners do not lose their
    separate identities.

17. A _____ is one
    that allows and encourages partners to grow
    and change, and spouses' roles may be
    renegotiated as the needs of each change.

Key Research Studies

You should be familiar with the main question
being investigated and the research results for
the following studies:

    Cuber and Harroff: utilitarian and intrinsic
        marriage
    Cuber and Harroff: five marriage styles
    Whyte: studies of wedding rituals

True-False

____ 1. The marriage premise refers to the
         assumption that marriage is no longer
         "until death parts us."

____ 2. Flexible marriages anticipate that one or
         both of the partners will change in time,
         and encourage or welcome such change.

____ 3. Widely accepted figures suggest that more
         than three-fourths of American husbands
         had at least one extramarital affair.

____ 4. The effect an extramarital affair has on
         a marriage is not always adverse or
         negative only.

____ 5. For many newlyweds, friendships with
         many of their single friends deteriorated
         after the marriage.

____ 6. "Egotism jealousy" occurs when one partner is expected by the other to conform to role stereotypes.

____ 7. A utilitarian marriage is one begun or maintained for primarily practical purposes.

____ 8. Intrinsic marriages are less vulnerable to divorce that are utilitarian marriages.

____ 9. Conflict-habituated marriages almost with exception end in divorce.

____ 10. Emotional emptiness does not necessarily threaten the stability of a marriage.

____ 11. Of the five types of marriage discussed by Cuber and Harroff, the MOST all-encompassing is the "vital" marriage.

____ 12. Exclusion jealousy occurs when the partner is afraid of being left out of the important aspects of a relationship.

____ 13. Partners in static marriages rely on their formal, legal bond to enforce permanence.

____ 14. Negotiating personal contracts helps to intensify the romanticism that is so important in forming sound engagements.

Multiple Choice

1. Rodolpho and Musetta are determined to put each other first in their lives, to try to make each other happy, and to try to make the marriage endure. Which of the following applies to this example?
    a. aggressive-conformist marriage
    b. the marriage premise
    c. the marriage gradient
    d. the N.A.S.H. process

2. The commitment of both partners to keeping each other most important person in their lives is called:
   a. fronting
   b. Christian love
   c. humanistic love
   d. primariness

3. All accepted figures suggest that _____ of all American husbands have had at least one extramarital affair.
   a. less than 10 percent
   b. less than 20 percent
   c. slightly less than 35 percent
   d. about 50 percent

4. Low self-esteem and fear of rejection by the partner is associated with:
   a. fear jealousy
   b. collegial jealousy
   c. monostatic jealousy
   d. neurotic jealousy

5. Which of these is NOT one of the types of jealousy discussed by Mazur?
   a. negotiation jealousy
   b. exclusion jealousy
   c. competition jealousy
   d. egotism jealousy

6. Newlyweds negotiate expectations for sex and intimacy, establish communication and decision-making patterns, and come to some agreement about childbearing. According to the text, they are involved in:
   a. creative institutionalization
   b. role-making
   c. love-nesting interaction
   d. primary institutionalization

7. Which of the following is most incompatible with "sexual exclusivity"?
   a. anticipatory socialization
   b. swinging
   c. professionalization
   d. social mobility

8. The intrinsic marriage offers:
   a. professional rewards
   b. community rewards
   c. external rewards
   d. intense emotional rewards

9. According to the text, _____ does not necessarily threaten the stability of a marriage.
   a. premarital pregnancy
   b. a homosexual affair
   c. emotional emptiness
   d. extreme economic instability

10. A "vital marriage" is one in which the partners:
    a. are brought together because of some pressing mutual crisis that they solve jointly
    b. marry because one of them has a pressing need for the marriage to occur
    c. marry because someone else wants the marriage to occur
    d. none of these

11. "Total" marriages are also:
    a. superficial
    b. devitalized
    c. on the increase
    d. intrinsic

12. According to the text, the nature and quality of a marital relationship has a great deal to do with:
    a. the political circumstances in which marital life must be lived
    b. the economic circumstances in which life must be lived
    c. the choices partners make
    d. the personal whim of the individual partners

13. A flexible marriage is one in which:
    a. a marriage contract or personal contract specifies what the requirements are
    b. people may change in terms of their sexual preferences
    c. the partners are not subject to the ordinary legal system and its definitions
    d. partners are relatively free to grow and change

14. According to the text, there is little evidence that couples who have lived together:
    a. have better marriages
    b. have higher levels of social mobility
    c. have lower levels of social mobility
    d. enjoy better overall physical and mental health

15. "Relationship agreements" differ from marriage contracts in that relationship agreements:
    a. specify emotional outcomes only
    b. specify practical outcomes only
    c. have an impact on emotional life, whereas marriage contracts have implications for behavior
    d. can apply to both married and unmarried couples

16. One important reason for writing a marriage agreement is that it helps partners to be aware of and avoid:
    a. ending up in closed marriages that are symbiotic
    b. choosing closed marriage by default
    c. the pitfalls of traditionalism in family and childbearing
    d. exorbitant attorney's fees associated with divorce and dissolution of marriage

17. Which of the following is explicitly mentioned by the text as an important reason for writing a marriage agreement?
    a. being specific about sex role expectations
    b. discovering the partner's past affectional attachments
    c. uncovering emotional disappointments that may affect the current relationship
    d. reducing the possibility of a failed relationship by 70-85 percent

18. According to the text, agreement with their spouses on the nature of the marital contract is significantly related to:
   a. the political liberality of the partners
   b. the higher level of partners' economic well-being
   c. positive motivations for an expected happy marriage relationship
   d. marriage adjustment

19. Constructing marital contracts has the positive consequence of helping partners to:
   a. enhance--not reduce--the euphoria of their love relationship
   b. communicate their social expectations to both sets of kin
   c. cut through or nullify the tendency to romanticize
   d. become more willing to accept the directional advice of third parties, whether kin or non-kin

20. According to the text, sometimes prenuptial contracts about property are:
   a. indicators of parental wishes more than of spousal desires
   b. disruptors of relationships that would have been successful had such contracts not been considered
   c. signed in a coercive atmosphere
   d. against state laws, as in most of the midwestern and southern states

---

Your Opinion, Please: What is your opinion about each of the following aphorisms or sayings?

"It is as absurd to say that a man can't love one woman all the time as it is to say that a violinist needs several violins to play the same piece of music." -- Balzac

"If you are afraid of loneliness, don't marry." -- Chekhov

Short-Answer Essay Questions
_____

The following are sample short-answer essay
questions--questions of the type you may asked if
your instructor uses questions like these. Even
if your instructor does not use questions like
these, you can help organize and consolidate your
learning if you can answer these questions in a
<u>well-organized</u> and <u>complete</u> manner. Do not think
that brief essays are easier than lengthy essays.
They may be more challenging because you are
required to be brief yet complete. And after you
have answered them, try making up some similar
questions and then answering them. Suggestion:
The Study Questions at the end of each text
chapter make very good short-answer essay
questions.

1. How have wedding ceremonies changed in recent
    years? In what ways are wedding ceremonies
    different for various population subgroups?

2. Summarize the effects of extramarital affairs
    on the relationship between married partners.

3. Distinguish between vital marriages and total
    marriages. Then distinguish between
    devitalized marriages and passive-congenial
    marriages.

<u>Essay</u>

The following are sample essay questions--
questions of the type you may be asked if your
instructor uses essay questions. Even if your
instructor does not use essay questions, you can
help organize and consolidate your learning if you
can answer these questions in a well-organized
and complete manner. Usually, the third essay
question is the most challenging.

1.   If there is "the marriage premise," when then
     are there extramarital affairs?

2.   Is it possible to make a contract for
     flexibility in marriage? Or is that a
     contradiction?

3.   In this chapter, the text explored at least
     seven different "types" of marriage. For
     example, utilitarian was discussed. Of the
     types discussed, select any four and fully
     explain each, perhaps constructing a
     hypothetical example to illustrate each.

<u>Answers</u>

Chapter Summary

the marriage **premise**
utilitarian and the **intrinsic** marriage
the **devitalized,** the passive-congenial
the **vital** and the **total** marriage
marriage or **domestic partner agreements**

Completion

1. utilitarian marriage
2. flexible marriage
3. primariness
4. sexual exclusivity
5. the marriage premise
6. primariness
7. conflict-habituated marriage
8. static marriage
9. primariness
10. utilitarian
11. passive-congenial
12. intrinsic
13. total marriage
14. devitalized marriage
15. personal marriage agreements
16. vital marriage
17. flexible marriage

True-False

1. F
2. T
3. F
4. T
5. T
6. T
7. T
8. F
9. F
10. T
11. F
12. T
13. T
14. F

Multiple Choice

1. b
2. d
3. d
4. a
5. a
6. b
7. b
8. d
9. c
10. d
11. d
12. c
13. d
14. a
15. d
16. b
17. a
18. d
19. c
20. c

CHAPTER 8
Communication and Conflict Resolution in Marriages and Families

---

Chapter Overview

This chapter examines some of the consequences of spouses' attempts to deny or ignore conflict and some more productive ways of coping with conflict. Among the results of such denial or refusal to face conflict are unhappiness, boredom, and fighting that only escalates the problems and results in more alienation. The chapter discusses some guidelines for fights that produce higher levels of bonding between partners and ways to change fighting habits. The chapter ends by discussing the myth of conflict-free conflict, communication, and family cohesion.

Chapter Summary

Families are powerful sources of support for individuals, and they reinforce members' sense of identity. Because the family is powerful, however, it can cause individuals to feel constrained, lacking a sense of individual freedom of expression and action. Tactics such as gaslighting, scapegoating, or negative use of the "_____ glass" can all be stressful or denigrating to an individual. On the other hand, couples may be able to counter the temptation to use anger "insteads"--passive-_____, _____-sacking, _____-sink fights--and avoid sending mixed (or _____) messages by using guidelines for bonding fights: listen, level with each other, use _____-statements, and give feedback; check your interpretation out, choose the time and place carefully, focus anger only on _____ issues, know what the fight is about, ask for change and be open to compromise, be willing to change yourself, and do not try to _____.
    Of course, changing fighting habits can involve both generational and couple changes, reflect gender differences, and reflect the types of relationships in which couples find themselves.

You should be able to explain the concepts listed
below. In your explanation, try to avoid using
the concept you are explaining. You should be
able to give several examples of each concept and
to explain <u>why</u> each example is an example.

looking-glass self
   (p. 259)
significant others
   (p. 259)
attribution
   (p. 259)
consensual
   validation
   (p. 262)
family cohesion
   (p. 277)
gaslighting
   (p. 262)
scapegoating
   (p. 262)
conflict taboo
   (p. 262)
anger "insteads"
   (p. 263)
displacement
   (p. 264)
alienating fight tactics
   (p. 266)

sabotage
   (p. 264)
gunnysacking
   (p. 268)
kitchen-sink fight
   (p. 268)
mixed, or double,
   message
   (p. 268)
bonding fighting
   (p. 269)
feedback
   (p. 271)
leveling
   (p. 270)
checking-it-out
   (p. 272)
traditionals
   (p. 276)
independents
   (p. 277)
separates
   (p. 277)

Point to Ponder: What do you think is the
                      "normal amount" that a
spouse may slap, hit, or beat his or her
partner? You may conclude that the normal
amount is none. Of course, not everyone in
every culture or subculture feels that way
about it. The fact that not everyone agrees
on this issue creates problems in a pluralistic
world in which people are supposed to be
allowed some cultural differences but are bound
by one common legal system.   How much, then,
should be allowed? None? Some? A lot?   And
if there are differences of opinion, how are
law enforcement agencies and courts to decide?

Complete the following sentences by selecting the correct alternative from the terms listed above. Some may be used more than once. Some may not be used at all. Filling in a blank may require more than one word.

1. The idea that conflict and anger are morally wrong and should be discouraged within the family is called the _____.

2. Persons whose opinions about each other are very important to each individual's self-esteem are referred to by sociologists as _____. Another way of expressing this is to say that these are persons whose standards are important to us as we evaluate our behavior.

3. When a person does not directly express anger to another but instead expresses it indirectly--often through a third person--then _____ is being used.

4. _____ fighting is the kind of fighting that brings people closer together rather than driving them farther apart.

5. The _____ refers to the process by which people come to see themselves as others see them.

6. _____ refers to those tactics that increase tension and conflict rather than reduce it.

7. When we assume that persons have certain character traits, we are engaging in the _____ process.

8. Paul is furious with Denise, but "nice husbands" don't slap their wives, so Paul kicks the cat. This can be interpreted as an example of _____.

9. The emotional bonding of family members is referred to as _____.

10. The idea of _____ refers to letting the other person know explicitly and completely exactly what you are feeling and thinking.

11. Comments such as "We're just not interested in making love anymore" or "No, I'd rather go bowling so I can't watch that program with you and your mother" can be interpreted as _____.

12. In _____ members of each group can come to some agreement about how each person sees the world. They may even come to some agreement about what reality is.

13. Andy plans a lawn party and asks Denise to arrange for the lawn tents. Denise "forgets" to do so and the lawn party is less than a success. This can be interpreted as an example of what the text terms _____.

14. To send a _____ is to send a contradictory message.

15. When one partner tries to change or distort the other partner's self-concept, the process is referred to as _____.

Key Research Studies

You should be familiar with the main question being investigated and the research results from the following studies:

Bach and Wyden: research on tactics for fighting
Stinnett: characteristics of strong families
Fitzpatrick: typology of marriage communication

Key Theories

Cooley: theory of the looking-glass self
symbolic interaction theory
systems theory

True-False

___ 1. Incorporating other people's responses about ourselves into our self-concept is part of what is meant by "internalization."

___ 2. Significant others are persons of relatively high status in the community whose marriages are often held up as role models for others to follow.

___ 3. Significant others are paid therapists to whom the couple with problems can turn for help.

___ 4. A person with a self-concept is a person who is more selfish than average.

___ 5. "You know how husbands are: late all the time, somewhat stingy, but handy around the house. You can't do with them and you can't do without them." This statement would serve well as an example of "attribution."

___ 6. Through consensual validation, family members can potentially help one another feel comfortable about how each perceives the world.

___ 7. Family cohesion is something that, from your text's point of view, is to be avoided insofar as is possible.

___ 8. A "spiritual orientation" is among the characteristics of strong families, according to Stinnett.

_____ 9. Married couples do not need to plan time for intimacy; they have it in the natural course of things. All they need to do is to take advantage of it.

_____ 10. "Gaslighting" is one of the techniques recommended by the text for creating a calm spousal argument.

_____ 11. Systems theory as discussed in the text examines the human capacity for sexual response as part of a general anatomical and psychological system.

_____ 12. A major criticism of systems theory was its vagueness.

_____ 13. The text presents the conflict taboo as something that should be eliminated.

_____ 14. Anger "insteads" refer to the text's suggestion that anger should be put aside and other activities be engaged in instead of behaving angrily.

_____ 15. In sabotage, one partner attempts to spoil or undermine some activity the other has planned.

_____ 16. Couples' conflicts are likelier to be resolved when they pretend that the conflicts do not exist. This lowers tension, which then increases communication between spouses, which aids in conflict resolution.

_____ 17. The text takes the point of view that as a general principle, fights should not be evaded.

_____ 18. Gunnysacking refers to one of the techniques for evading a fight temporarily.

_____ 19. The test's recommended rules to govern fighting suggest that people avoid use of such phrases as "I feel...," "I think...," and so on.

_____ 20. Not every conflict can be resolved.

Multiple Choice

1. The looking-glass self is:
    a. the process whereby people come to see
        themselves as others see them
    b. a narcissistic phenomenon in which people
        have an unrealistic view of themselves
    c. a process whereby people look at
        themselves and have a conflict about the
        direction in which they should go
    d. the self one sees in the mirror "the
        morning after" a serious fight with the
        partner

2. Role taking refers to the process whereby a
    person:
    a. listens to and follows the advice of a
        marital therapist
    b. tries to connect his/her roles to some of
        the statuses offered by society
    c. plays out the expected behavior associated
        with a social position
    d. misappropriates for himself/herself the
        social roles of the partner or of another
        person in the family

3. Attribution is:
    a. the ascribing of certain character traits
        to persons
    b. feeling that a man or woman has many of
        the physical attributes that society says
        make a person attractive
    c. the process whereby the older family
        members are taken away by death
    d. one of the six methods mentioned by
        Martinelli as "fair fighting techniques"

4. Family members' viewpoints are an essential
    part of:
    a. the kinship negotiation process
    b. determining the rules by which we must
        live
    c. consensual validation
    d. interpersonal "U-turns"

5. Focusing on one family member to blame for almost everything that goes wrong in a family is what is meant by:
   a. gaslighting
   b. passive-aggression
   c. mixed message
   d. scapegoating

6. Gaslighting is best exemplified by which of the following remarks from a husband to his wife?
   a. "I personally prefer natural light of candles to artificial light from bulbs."
   b. "Every time you balance the checkbook, you make math errors. Our accountant says the errors are in **your** handwriting."
   c. "Who turned the television on? <u>You</u> turned it on. You must be losing your mind."
   d. "Don't turn the ceiling lights on. It's much more pleasant with the light coming from the fireplace. You look even more handsome by firelight."

7. A basic idea in _____ is that of _____.
   a. interaction; functions
   b. functions; symbolic interaction
   c. symbolic interactionism; imaging
   d. systems; feedback

8. Which of the following is closely associated with the practice of therapy?
   a. functionalism
   b. cognitive dissonance
   c. systems theory
   d. conflict structuralism

9. Conflict taboo refers to which of these?
   a. jealous arguments started as a result of the wearing of a famous perfume
   b. the idea that taboos can create conflicts between partners if the taboos are not openly discussed
   c. the concept that close kin become political adversaries when there is a conflict
   d. none of these

10. Passive-aggression is a concept that refers to which of these?
    a. the feeling that one's spouse is being both passive and aggressive
    b. the view that one should be passive in the face of aggression
    c. the view that one should not be passive in the face of aggression
    d. none of these

11. Turning sullen and refusing to talk, or stating "I can't take you seriously when you act this way," are examples of:
    a. marriage inflexibility
    b. denial
    c. self-actualization
    d. fight evading

12. In fights, a rule to use in avoiding attack is to use _____ instead of _____.
    a. "sometime"; "never"
    b. gender-neutral words; gender-specific words
    c. actions; words
    d. "I"; "you" or "why"

13. A "kitchen-sink fight" is one in which:
    a. the combatants use any argument available to them, appropriate or not
    b. the couple fights at breakfast
    c. the couple fights in the kitchen, regardless of the time of day
    d. the "dirtiest dishes" in the couple's history are the topic for discussion during a family argument

14. Which of these is a method recommended by the text to avoid attacks on a partner's self-esteem?
    a. Use "I-statements."
    b. Avoid topics in which the partner's self-esteem is an issue.
    c. Avoid topics that could have the effect of damaging the partner's self-esteem
    d. Take as much of the blame as is possible on oneself.

15. Feedback exists when one of the partners:
    a. delivers a reply to the partner that is as devastating as the ones received from the partner
    b. repeats the partner's criticism but applies it to the partner instead of to oneself
    c. repeats in one's own words what the other partner has said or revealed
    d. uses the present argument to recall and draw on previous arguments

16. Which of the following is one of the guidelines for bonding fights?
    a. Don't try to win.
    b. Avoid giving negative information to the partner.
    c. Try to present yourself as you would would like to be, not as you think you are.
    d. Fights should always end with an agreement between partners.

17. According to the text, it is a myth that there is any such thing as conflict-free ___.
    a. love affairs
    b. primary friendship
    c. amicable love
    d. conflict

18. A study comparing mutually satisfied couples with those experiencing marital difficulties found that when couples are having trouble getting along or are under "stressed" conditions, they tend to:
    a. increase communication with same-sex friends
    b. interpret each other's messages and behavior more negatively
    c. internalize the problems and reduce their amount of communication
    d. withdraw into themselves more and emphasize work or other projects

19. One characteristic of happier couples is:
    a. similar levels of language skills
    b. a healthy preference for keeping negative or unpleasant thoughts to oneself
    c. the restraint each uses with regard to expressing negative thoughts and feelings
    d. the ability to freely express any thoughts or points of view, however positive or negative

20. According to the text, it is important to know why one wants to tell negative information to a partner (to "win"?) and whether the partner really needs to know the information. For example, the text uses which of the following to illustrate this principle?
    a. an overdrawn checking account
    b. a past extramarital affair
    c. personal concern about the results of one's own medical checkup
    d. information that the whereabouts of one's daughter/son is unknown, when this is not yet a certainty

21. According to George Bach, who founded and directs an institute to teach "constructive aggression," fight training:
    a. does not always end happily; it can even end in divorce
    b. works best when the partners can be characterized as "traditional" couples
    c. is a concept that should probably be eliminated from the vocabulary of human service professionals
    d. should be part of the socialization of every person, beginning about age 4

22. Citing research, the text states that the answer to the question about which came first and which later, the answer is clearly that _____ comes first, followed by _____.,
    a. poor dating relationships; poor marriage relationships
    b. fighting; marital distress
    c. thoughts about divorce; fighting
    d. poor communication skills; marital distress

23. The textbook states that the novelist Leo Tolstoy was incorrect in saying _____, because evidence indicates to the contrary.
    a. husbands' pains are wives' joys
    b. husbands are less "naturally" husbands than are wives "naturally" wives
    c. husbands communicate best their thoughts; wives communicate best their feelings
    d. happy marriages are all alike; unhappy marriages are infinitely varied

24. According to research by Fitzpatrick, the more partners' _____ matched their own ideologies of marriage, the more satisfied the couple was with their marriage.
    a. truce-making skills
    b. communication
    c. rules for the distribution of "fall-out from fighting"
    d. ideologies of individualism

25. Which of the following is not one of Fitzpatrick's couples' marriage ideologies?
    a. democratic
    b. traditionals
    c. independents
    d. separates

26. Sally and Billy have conventional beliefs about marriage and are unhappy with conflict avoidance on major problems, readily addressing serious issues. They illustrate which of these marriage ideologies?
    a. democratic
    b. traditional
    c. independents
    d. separates

27. Not very interdependent, having conservative sex role beliefs, and "emotionally divorced" is which of the following types of marriage ideologies?
    a. mixed
    b. traditional
    c. independents
    d. separates

28. When Fitzpatrick administered the "Dyadic Adjustment Scale" to couples, the type receiving the highest adjustment scores were:
    a. mixed
    b. traditionals
    c. independents
    d. separates

29. Stinnett studied observations about family strengths and found that _____ qualities stood out in strong families.
    a. five
    b. six
    c. seven
    d. eight

30. Stinnett found that strong families had all of the following characteristics EXCEPT which one?
    a. deal positively with crises
    b. positive communication patterns
    c. a spiritual orientation
    d. the personal integrity to establish a reasonable individual schedule and stick to it

---

Your Opinion, Please: Some people feel that there are times when you are so angry that it is best to keep your thoughts and feelings to yourself. When something is said, it can never be taken back, the slate cannot be wiped clean, and the relationship can never be the same again. Others think that you are entitled to your feelings and beliefs, and that you should let the other person know how you think or feel about things, no matter what the consequences. Has someone ever asked you for your honest opinion and, when you gave it, attacked you for your opinion. If it happened, how did it make you feel about that situation? About more communication with that person?

Short-Answer Essay Questions

The following are sample short-answer essay questions--questions of the type you may be asked if your instructor uses questions like these. Even if your instructor does not use questions like these, you can help organize and consolidate your learning if you can answer these questions in a <u>well-organized</u> and <u>complete</u> manner. Do not think that brief essays are easier than lengthy essays. They can be more challenging because you are required to be brief yet complete. And after you have answered these, try making up some similar questions and then answering them. Suggestion: The Study Questions at the end of each text chapter make very good short-answer essay questions.

1. Briefly explain each of the following: anger "insteads" and "kitchen-sink" fights.

2. Explain the difference between "I-statements" and leveling with each other.

3. Distinguish between "independents" and "separates" as styles of communication among couples.

Essay

The following are sample essay questions--
questions of the type you may be asked if your
instructor uses essay questions. Even if your
instructor does not use essay questions. you can
help organize and consolidate your learning if you
can answer these questions in a well-organized and
complete manner. Usually, the third essay
question is the most challenging.

1. What is the connection between communication
    in families and symbolic interactionism?

2. Explain the various alienating practices
    explored in the text. For each, give an
    example of your own devising, and in your
    answer make it clear why each example is a
    good example.

3. Explore the origins and consequences of
    conflict in families. Where appropriate, back
    up your essay with appropriate specific
    concepts and research results.

Answers

Chapter Summary

the "looking glass"
passive- aggressive
gunny- sacking and kitchen- sink
mixed (or double) messages
use I- statements
only on specific issues
not trying to win

Completion

1. conflict taboo
2. significant others
3. passive-aggression
4. bonding
5. looking-glass self
6. alienating fighting
7. attribution
8. displacement
9. family cohesion
10. leveling
11. anger "insteads"
12. consensual validation
13. sabotage
14. mixed or double message
15. gaslighting

True-False

1. T
2. F
3. F
4. F
5. T
6. T
7. F
8. T
9. F
10. F
11. F
12. T
13. T
14. F
15. T
16. F
17. T
18. T
19. F
20. T

Multiple Choice

1. a
2. c
3. a
4. c
5. d
6. c
7. d
8. c
9. d
10. d
11. d
12. d
13. a
14. a
15. c
16. a
17. d
18. b
19. c
20. a
21. a
22. d
23. d
24. b
25. a
26. b
27. d
28. b
29. b
30. d

CHAPTER 9
Power in Marriage and Families

Chapter Overview

This chapter examines power relationships in marriage and families. The main focus of attention is what sociologists call "conjugal" power and decision making. The text takes the point of view that playing power politics in marriage is harmful to intimacy. Alternatives to power politics are explored.

Chapter Summary

Power is the ability to exercise one's will and may rest on cultural authority, economic and personal resources that are gender based, love and emotional dependence, interpersonal manipulation, or physical violence.
   The relative power of a husband and wife in a marriage varies by national background and race, religion, and class. Blood and Wolfe call this the _____ hypothesis. Such power varies by whether or not the wife works and the presence and age of children. American marriage experiences a tension between male dominance and egalitarianism. Studies of married couples, cohabiting couples, and gay male and lesbian couples illustrate the significance of economic-based power, as well as the possibility for couples to consciously work toward more egalitarian relationships.
   Physical violence is the most commonly used in the absence of other resources. While men and women are equally likely to abuse their spouses, the circumstances and outcomes of marital violence indicate that wife abuse is a more crucial social problem. It has received the most programmatic attention. Recently, programs have been developed for male abusers, but less attention has been paid to male victims. Experiments indicating that arrest is an important deterrent to further wife abuse illustrate the importance of public policies to meet family needs.

Economic hardships and concerns (for parents of **all** social classes and races) can lead to physical and/or emotional child abuse--a serious problem in our society and probably far more common than statistics indicate. One difficulty is drawing a clear distinction between "normal" child rearing and abuse.

Elder abuse and neglect is a new area of research, but early data suggest that abused elderly are often financially independent and abused by dependent adult children or by elderly spouses.

While some scholars view elder abuse and neglect as primarily a care giving issue, this chapter presents elder abuse as a family violence and family power issue.

Key Terms

You should be able to explain the concepts listed below. In your explanation, try to avoid using the concept you are explaining. You should be able to give several examples of each concept and to explain why each example is an example.

conjugal power
   (p. 284)
personal power
   (p. 284)
power
   (p. 284)
social power
   (p. 284)
resource hypothesis
   (p. 284)
principle of least
   interest
   (p. 293)
relative love and
   need theory
   (p. 293)
no-power
   (p. 297)
power politics
   (p. 297)

neutralizing power
   (p. 298)
battered woman syndrome
   (p. 307)
child abuse
   (p. 312)
sexual abuse
   (p. 312)
incest
   (p. 312)
child neglect
   (p. 312)
elder abuse
   and neglect
   (p. 315)
marital rape
   (p. 306)
three-phase cycle of
   violence
   (p. 306)

Completion

Complete the following sentences by selecting the
correct alternative from the terms listed above.
Some may be used more than once. Some may not be
used at all.   Filling in a blank may require more
than one word.

1.   _____ can be defined as the
      ability to exercise one's will (or, to get
      one's way.

2.   In a _____, each partner
      knows vicariously how to play the other's
      role, that is, knows how to play either the
      dominant or the submissive role.

3.   The _____ refers to
      the fact that the partner who has less to
      lose from ending the relationship is the one
      who is more apt to exploit the other partner.

4.   When a subordinate weakens the powerful
      person's control by refusing to cooperate in
      that power, the term _____
      is applicable.

5.   Power exercised over oneself is called
      _____.

6. Blood and Wolfe in their _____
   argued that power in marriage comes mainly
   from the adequacy of the various things one
   can draw on or "fall back on" as leverage to
   "get one's way."

7. Power between married persons is termed
   _____.

8. _____ means that both
   partners wield about equal power.

Key Research Studies

You should be familiar with the main question
being investigated and the research results for
the following studies:

   Blood and Wolfe: decision-making structures
      in families
   Straus et al.: incidence of family violence
      among husbands, wives, and cohabitors
   Blumstein and Schwartz: American couples:
      money, work, and sex
   National Family Violence surveys: 1975 and
      1985

Key Theories

   Blood and Wolfe: resource hypothesis
   Waller: principle of least interest
   relative love and need theory
   The caregiving model of family violence and
      the domestic violence model of family
      violence

True-False

____ 1. Power that an individual exercises over
         others is "social power."

142

____ 2. Power between married partners is referred to as conjugal power.

____ 3. According to the resource hypothesis, the spouse with more resources has more power in marriage.

____ 4. Researchers have found that the higher the education and occupational status of husbands, the greater their conjugal power.

____ 5. The resource hypothesis tends to focus narrowly on partners' individual personalities and the way in which these personalities interact.

____ 6. American marriages tend to be inegalitarian.

____ 7. In our society, we have an ideology of marital equality, but male dominance continues to prevail.

____ 8. It may be a myth that patterns of power among blacks and Mexican-Americans are very different that those among whites.

____ 9. The principle of least interest refers to the tendency of school boards to elect as administrative disciplinarian one whom students find both boring and fear-inspiring.

____ 10. The relative love and need theory assumes that women are as likely to have power as men are.

____ 11. The relative love and need theory predicts that husbands will generally be more powerful.

____ 12. Historically, women have tended to rely on "micromanipulation."

_____ 13. Lipen-Blumen argues that men dominate the public spheres of work and political leadership but that women tend to dominate the private sphere.

_____ 14. A recent study by Blumenstein and Schwartz of married heterosexual couples, cohabiting couples, and lesbian couples found that gender was by far the most significant determinant of the pattern of power in these relationships.

_____ 15. Marital violence exists in all social classes.

_____ 16. More programs/services exist to serve the needs of gays to cope with domestic violence than exist for other groups.

_____ 17. Marriage counselors are for the most part committed to helping couples avoid no-power relationships.

_____ 18. Wife-beating occurs more often in marriages where the wife can express herself better than can her husband.

_____ 19. No marriage is entirely free of power politics.

_____ 20. Battered wives' lack of personal power begins with their faith in violent power.

Multiple Choice

1. In the purest sense (as explored by the text), "power" is:
   a. the ability to require people to do what you want them to do
   b. something that is granted by the norms of religion
   c. something that is granted by the legal norms
   d. roughly equivalent to "influence"

2. Social power is the kind of power that an
    individual exercises:
    a. among the upper-middle and upper classes
    b. over oneself, and is similar to self-
        control
    c. over oneself in public, in accordance
        with social norms
    d. over others

3. Research by Blood and Wolfe found that the
    more _____ one has, the more power one has
    over making decisions in families.
    a. will power
    b. flexibility
    c. futurity
    d. resources

4. In their research, Blood and Wolfe found that
    _____ husbands have more final decisions
    than _____ husbands.
    a. present; absent
    b. deviant; innovative
    c. innovative; conformist
    d. white-collar; blue-collar

5. A serious criticism of the Blood and Wolfe
    research is that these authors did not take
    into account:
    a. social class
    b. resources
    c. education
    d. personality

6. The resource hypothesis presents resources as
    _____ and power as _____.
    a. problematic; inevitable
    b. inevitable; gender-based
    c. a variable; a constant
    d. neutral; gender-free

7. According to the text, _____ is a
    myth.
    a. black matriarchy
    b. white patrimony
    c. Mexican-American avuncularity
    d. minority group endogamy

8. The marriages of blacks tend to be more ____
    than those of whites.
    a. free-form
    b. egalitarian
    c. matriarchal
    d. innovative

9. According to the text, "micropolitics" is
    something that goes on:
    a. within families
    b. within one's own mind
    c. within neighborhoods
    d. within the imaginings of "the Beloved"

10. As the text uses the term, _____ seek to
    negotiate and compromise, not to "win."
    a. segmental power couples
    b. no-power couples
    c. patriarchal couples
    d. autocratic couples

11. Which of these means that both partners exert
    about equal power?
    a. no-power
    b. quasi-equilibrated power
    c. centered power
    d. scattered power

12. One study of college students found that
    _____ couples were least happily married.
    a. religiously out-married
    b. religiously in-married
    c. exchange-oriented
    d. tradition-oriented

13. In "neutralizing power," the subordinate
    person:
    a. has counterarguments
    b. refuses to cooperate, takes a neutral
        position
    c. matches blow for blow and punch for punch
        with the opposition
    d. quotes past happenings in the
        relationship, which makes the opponent
        withdraw from arguing

14. An estimated ____ of murders of women by
    their male partners occurred in response to
    the woman's attempt to leave.
    a. 30 percent
    b. 47 percent
    c. 55 percent
    d. 75 percent

15. Together, the 1975 and 1985 surveys found
    that in ___ percent of the couples, at least
    one of the partners had engaged in a violent
    act against the other during the previous
    year.
    a. 16
    b. 27
    c. 38
    d. 65

16. According to the text's Table 9.1, the type
    of family violence with the **highest** rate of
    occurrence is:
    a. violence between husband and wife
    b. violence by parents against a child under
        age 18
    c. violence by parents against a child age
        15-17
    d. violence by one sibling against another
        sibling

17. Wife abuse results in serious injuries. A
    survey of physical damage typical of battered
    women surveyed in the early 1980s in
    California shelters found that most injuries
    were injuries to:
    a. the head and neck
    b. the upper back and/or the chest area
    c. the mid-torso
    d. the extremities--usually to forearms

18. A 1980-81 Boston study reported that about
    ____ of wives' husbands used physical force
    or threats to compel sex.
    a. 10 percent
    b. 24 percent
    c. 38 percent
    d. 52 percent

19. Husbands can now be accused of marital rape in _____ of the states, at least while spouses are living together.
    a. just under one-fourth
    b. approximately one-third
    c. slightly over one-half
    d. nearly four-fifths

20. In the three-phase cycle of domestic violence, the first phase is:
    a. the wife and husband exist in harmony
    b. the wife taunts or pushes her husband too far and he strikes out against her
    c. tension builds up over some minor altercations over a period of time
    d. the husband is shocked when the mate reports the violence to legal authorities

21. In the three-phase cycle of domestic violence, the second phase consists of:
    a. a period of domestic calm
    b. an escalating series of tension-evocative situations
    c. the husband commits a violent act against the wife
    d. the husband is apologetic and contrite

22. Why do abusive husbands do it? Studies indicate that husbands who beat their wives are attempting to:
    a. compensate for general feelings of powerlessness
    b. copy the relationship skills in their own families of origin
    c. copy the relationship skills they see in the visual media, especially television
    d. socialize their wives to expected modes of domestic behavior, using force as a means of last resort

23. A study of blue-collar families found that husbands who _____ were likeliest to beat their wives.
    a. had more rather than fewer children
    b. had histories of frequent formal contact with the police
    c. were in second or third marriages
    d. did not have superior legitimate resources to maintain dominance

148

24. A cross-cultural analysis of domestic violence in 90 different societies found family violence to be virtually absent in 16 societies. Among the characteristics of these 16 societies was:
   a. living in the midst of abundant natural resources
   b. a hierarchical or pyramidal power structure in society
   c. communal child rearing or parenting by both sexes
   d. economic and decision-making equality between the sexes

25. Kalmuss and Straus found in a national sample that "severe violence" was associated with women's:
   a. economic dependence as measured by employment status
   b. inappropriate psychological and interactional assertiveness over men
   c. inability to express themselves verbally or in terms of gestures, as measured using the Bales Interaction Recorder
   d. infidelity or occurrence of pregnancy known to be unwanted by the male partner

26. Kalmuss and Straus did **not** find that _____ was related to women experiencing severe, violent spouse abuse.
   a. subjective marital dependence
   b. objective marital dependence
   c. subjective psychological individualism
   d. objective psychological individualism

27. According to the text, unlike husbands' violent acts, the violent acts of wives:
   a. do not tend to be highly symbolic of personal feelings
   b. do not tend to be repeated over time
   c. tend to be displaced away from the husband and directed to household objects
   d. tend to be inner-directed or directed toward the self, as in cases of suicide

28. The text cites one estimate that about _____
    of victimized women will be revictimized
    within a relatively short time unless there
    is effective intervention.
    a. 21 percent
    b. 32 percent
    c. 47 percent
    d. 68 percent

29. Research indicates that:
    a. wives' violence against husbands is less
        frequent, but it tends to be more violent
    b. much of husbands' violence appears to be
        in self-defense against mildly abusive
        wives
    c. husbands are less apt than are wives to
        leave an abusive relationship in a short
        period of time
    d. wife abuse tends to escalate, growing
        more severe over time

30. Data from one study of 300 cases of the abuse
    of elderly persons in the Northeast found
    that abusers (frequently an adult son), were
    likely to be:
    a. emotionally overextended as a result of
        intensive caregiving to the abused
        elderly person
    b. without remorse for their abusive
        behavior, almost as if the elderly had no
        capacity to understand the abusive
        behavior
    c. financially dependent on the elderly
        victim
    d. younger rather than older caregivers

Answers

Chapter Summary

the **resource** hypothesis

Completion

1. power
2. no-power situation
3. principle of
   least interest
4. neutralizing power
5. personal power
6. resource hypothesis
7. conjugal power
8. no-power

True-False

1. T
2. T
3. T
4. T
5. F
6. T
7. T
8. T
9. F
10. T
11. F
12. T
13. T
14. T
15. T
16. F
17. F
18. T
19. T
20. T

Multiple Choice

1. a
2. d
3. d
4. d
5. d
6. d
7. a
8. b
9. a
10. b
11. a
12. c
13. b
14. d
15. a
16. d
17. a
18. a
19. c
20. c
21. c
22. a
23. d
24. d
25. a
26. a
27. b
28. b
29. d
30. c

CHAPTER 10
To Parent or Not to Parent

Chapter Overview

This chapter examines the choices individuals and couples are making about whether to have children, the pressures and responsibilities that accompany this choice, and the methods available to help people control family size. Scientific and technological advances have both increased people's options and added new concerns to their decision making. But technological progress does not mean that people can or do exercise complete control over their fertility. Options have increased, but some things remain beyond the individual's or couple's control.

Chapter Summary

Today, individuals have more choice than ever about whether, when, and how many children to have. Although parenthood has become more of an option, there is no evidence of an embracement of childlessness. The majority of Americans continue to value parenthood, believe that childbearing should accompany marriage, and feel social pressure to have children--a _____ bias in our society. Only a very small percentage view childlessness as an advantage, regard the decision not to have children as positive, believe that the ideal family is one without children, or expect to be childless by choice.

Nevertheless, it is likely that changing values concerning parenthood, the weakening of social norms prescribing marriage and parenthood, a wider range of alternatives for women, the desire to postpone marriage and childbearing, and the availability of modern contraceptives and legal abortion will eventually result in a higher proportion of Americans remaining childless. In fact, some observers have begun to worry that American society may be drifting into a period of _____ antinatalism.

Children can add a fulfilling and highly rewarding experience to people's lives, but they also impose complications and stresses, both financial and emotional. Couples today are faced with options other than the traditional family of two or more children: remaining childless, postponing parenthood until they are ready, and having only one child. Often, people's decisions concerning having a family are not made knowledgeably but instead are made by _____.

Birthrates are declining for married woman. White women especially are waiting longer to have their first child and are having, on the average, two children. Although pregnancy outside of marriage has increased, many unmarried pregnant women choose abortion.

For couples who have difficulty in conceiving, reproductive technologies include artificial insemination, _____ fertilization, embryo transplant, and the use of a _____ mother. There are many social and ethical issues surrounding these procedures. Also, adoption is a way of becoming a parent without conceiving; some families have both adopted and biological children.

Key Terms

You should be able to explain the concepts listed below. In your explanation, try to avoid using the concept you are explaining. You should be able to give several examples of each concept and to explain why each example is an example.

crude birthrate
   (p. 324)
fecundity
   (p. 325)
fertility
   (p. 325)
total fertility rate
   (p. 324)
pronatalist bias
   (p. 329)
structural antinatalism
   (p. 330)

opportunity costs
   (p. 332)
programmatic postponers
   (p. 325)
paradoxical pregnancy
   (p. 348)
abortion
   (p. 349)
involuntary infertility
   (p. 354)
subfecundity
   (p. 354)

Point to Ponder: Every society faces this difficult issue: How to replace people who die and so have enough people to do what needs to be done to keep society going. Most societies achieve this by sexual reproduction or by immigration. But some societies have found they are under-reproducing their population. The birthrate has fallen below the replacement level: more people are dying or leaving than are being born or arriving as immigrants. Examples include the Island of Malta and the nation of Hungary. How long can a society afford to underrepro-duce? And what can be done to make it possible for society to survive under such conditions? Since fertility rates are sensitive to uncertainties and to economic conditions, does this suggest a role for government (tax breaks for having children, etc.)?

Completion

Complete the following sentences by selecting the correct alternative from the terms listed above. Some may be used more than once. Some may not be used at all. Filling in a blank may require more than one word.

1. When various aspects of culture and society have an institutionalized bias against having children, or when parents are penalized in some ways for having children, this is referred to as _____.

2. When aspects of culture and society encourage people to avoid contraception and to have children, then it is appropriate to speak of _____.

3. The _____ has been declining for over 100 years except for the baby-boom period of the 1940s and 1950s.

4. The term _____ refers to reduced reproductive ability.

5. According to the text, couples choose _____ far more often than people may realize.

6. _____ is the expulsion of the fetus or embryo from the uterus either naturally or medically.

7. When parents have to give up income and investments because they have decided to rear children, then it is appropriate to refer to this as _____.

8. With a(n) _____, a fertilized egg is implanted into an infertile woman.

9. Partners who set a schedule of personal goals, occupational plans, and educational efforts and intentionally build childbearing in that scheduled are called _____.

10. The term _____refers
    to the fact that the more guilty and
    disapproving women are about premarital sex,
    the less likely they are to use
    contraceptives regularly, if at all.

11. With _____ a baby
    is conceived outside the woman's body but
    develops within a woman's uterus.

12. Couples who want to conceive and give birth
    to a child but cannot physically do so
    experience what the text terms
    _____.

13. According to the text, white women are more
    likely to engage in _____
    through formal procedures than are African-
    American women or Latinas.

14. _____ refers to the
    biological capacity or ability to reproduce.

15. In _____, biological
    and adoptive families exchange personal
    information, such as letters or photographs,
    but do not have direct contact.

_____

True-False

____ 1. The total fertility rate in the United
        States is approximately 1.9 at present.

____ 2. The current trend in declining American
        fertility is a continuation of a long-
        term pattern dating back to about 1800.

____ 3. It is only when individuals have
        satisfying options other than parenthood
        that they choose to limit their
        childbearing.

____ 4. Among Hispanics, the fertility rate for
        Puerto Rican women is lower than for
        Mexican-American women.

_____ 5. Pronatalists are a group with a strong likelihood of joining a voluntary childlessness group because pronatalist have the same basic goals and attitudes as those who endorse voluntary childlessness.

_____ 6. Structural antinatalism refers to aspects of personality structure that result in a lower fertility rate.

_____ 7. Children are the only age group over-represented in the poverty population.

_____ 8. Opportunity costs are the expenses to be borne by those wanting to maximize their geographical placement and general attractiveness (activities, leisure, appearance) so as to meet and marry a desirable spouse.

_____ 9. The costs of having children include "opportunity costs."

_____ 10. The text states that marital happiness tends to be higher in child-free unions.

_____ 11. About 50 percent of couples report satisfaction with their decision to become parents.

_____ 12. Child-free women do not tend to be attached to a satisfying career.

_____ 13. There are no major differences between "only" children and children who have one or more siblings.

_____ 14. Unmarried pregnancy is twice as likely to happen among blacks as it is to happen among whites.

_____ 15. Complications of pregnancy, miscarriage and stillbirth, prematurity, and birth defects are more likely with teenage mothers than with mothers in their 20s and early 30s.

_____ 16. "Paradoxical pregnancy" refers to a woman who simultaneously wants to have a baby AND wants to conform to traditional norms about premarital chastity.

_____ 17. From earliest history, abortion has been a way of preventing birth.

_____ 18. The "Roe v. Wade" decision has restricted federal funding for abortions, permitted states to require notification of minors' parents, and permitted states to require waiting periods before abortions.

_____ 19. A "hysterotomy" is a surgical procedure whereby the woman's abdomen is opened and both the uterus and the fetus are removed, ending the pregnancy and permanently sterilizing the woman.

_____ 20. According to footnote #10, a survey of married women who obtained abortions found that almost 90 percent of them had told their husbands.

Multiple Choice

1. The total fertility rate in the United States reached an all-time high in:
   a. 1935
   b. 1946
   c. 1957
   d. 1965

2. The highest recorded total fertility rate in the United States was which of these?
   a. 3.6 births
   b. 0.78 births
   c. 6.1 births
   d. 120 births

3. Today, women of all ages want a family consisting of which of how many children?
   a. two sons
   b. two daughters
   c. two children, of whichever sex
   d. three children

4. According to the text, the birthrate for blacks in this country did not decline significantly until _____, when it began to decline.
   a. 1880
   b. 1910
   c. 1936
   d. 1966

5. Which of these is the OPPOSITE of "pronatalist"?
   a. being for early marriage
   b. being against contraception
   c. supporting couples being child-free or childless
   d. affirming cohabitation and group marriage among the elderly

6. Lower-income parents:
   a. are likely to have antinatalist sentiments and values
   b. are less likely to use abortion than are middle-class persons
   c. are among the strongest supporters of having high-tech babies
   d. may perceive rewards in having children that they may not get any other way

7. The possibilities of more wage earning and investments that parents frequently give up to have and rear children is:
   a. usually recovered at the other end of their life span
   b. relatively negligible until the offspring are in junior high school and/or later
   c. ordinarily most consequential in the children's earliest years, especially in an economic sense
   d. referred to as "opportunity cost"

8. Statistical evidence shows that children do which of the following?
   a. increase feelings of romantic love between the spouses, especially during preschool years
   b. motivate parents toward higher occupational aspiration and more creative problem solving at work
   c. make "shaky" marriages better, socially and psychologically, between the spouses
   d. stabilize marriages

9. When a husband and wife disagree about having children, the wife usually:
   a. wins the argument
   b. discusses the issue and a democratic decision is usually reached
   c. begins having "medical problems" as a way out of the situation
   d. accepts her husband's position on the matter

10. Studies comparing the characteristics of "only" children with the characteristics of children with one or more siblings find that the "only" children tend to be superior in which of the following ways?
    a. they tend to have "perfect" pitch in music
    b. they tend to have better skills at rote memorization
    c. they tend to have more self-reliance and self-confidence
    d. they tend to have lower frustration levels

11. According to Census Bureau reports, by 1988 about _____ of all births occurred among unmarried mothers.
    a. 5 percent
    b. 10 percent
    c. 17 percent
    d. 26 percent

12. Glenn and McLanahan found that the only category of people who thought there were more positive than negative effects when there were children in the home were which of these?
   a. highly educated black people with two children
   b. highly educated white people with two children
   c. lowly educated white people with no children yet
   d. white people agreeing that four or more children at home is a desirable thing

13. The text reports one study of only children in which only children were found to have _____ when compared to children with siblings.
   a. greater capacity for concentration
   b. less tolerance of others' eccentricities
   c. greater empathy or ability to take others' point of view into account
   d. greater verbal skills

14. Unwed birthrates have increased dramatically for _____ in recent years.
   a. females reporting no religious affiliation
   b. Protestants, especially Southern Baptist women
   c. older women
   d. the upwardly mobile

15. According to the text, there is _____ to future pregnancies from a first-trimester vacuum aspiration abortion of a first pregnancy.
   a. no risk
   b. a 15 percent risk
   c. a 26 percent risk
   d. a 35 percent risk

16. Amniocentesis is:
    a. inability through "repression" for a woman to remember her recent abortion
    b. a term referring to the human instinct to want to have a child
    c. an abortion method for a woman who has become pregnant through rape
    d. a type of prenatal test

17. Artificial insemination is:
    a. a medical procedure that may make a woman successful in her desire to become pregnant
    b. a surgical procedure to discover physical barriers to successful pregnancy
    c. a way of fooling the body into believing that the sperm is that of a woman's husband rather than that of someone else
    d. a way to improve ovulation rate through proper diet

18. The acronym IVF means:
    a. involuntary fatherhood
    b. inactive vesicular freezing
    c. innovative viral fractioning
    d. in vitro fertilization

19. In artificial insemination, which of the following ordinarily occurs?
    a. a physician injects live sperm into a woman's vagina when she is ovulating
    b. artificial sperm "trick" the woman's ovum into fertilizing itself
    c. sperm are introduced into an artificial ovum that has been constructed to maximize fertility
    d. the physician uses an artificial device to stimulate the woman to orgasm, which is ordinarily sufficient to guarantee fertilization

20. About what percent of adoptions today are interracial?
    a. 58 percent
    b. 42 percent
    c. 27 percent
    d. 8 percent

21. The National Committee for Adoption estimates that about _____ children are legally adopted by nonrelatives (e.g., not including stepparents) each year.
    a.   50,000
    b.   150,000
    c.   550,000
    d.   900,000

22. According to the text, _____ are likelier to adopt through formal adoption procedures than are the other women listed below.
    a.   white women
    b.   black women
    c.   Latinas or Hispanic women
    d.   none of the above

23. Adoptions increased steadily through much of this century and reached a peak in _____.
    a.   1945
    b.   1970
    c.   1980
    d.   1991

24. Unmarried mothers are _____ to keep their babies.
    a.   increasingly likely
    b.   neither more nor less likely
    c.   less likely
    d.   much less likely

25. According to Table 10.2 in the text, in 1982-1988, among children born to never-married women, the percentage who were relinquished for adoption was:
    a.   0.45 percent
    b.   0.10 percent
    c.   2.0 percent
    d.   5.5 percent

26. According to Box 10.3 in the text, Deciding to Relinquish a Baby, the general conclusion of the research results supports the idea that relinquishers:
    a. experience a great deal of dissatisfaction with their decision to place the baby for adoption
    b. do not experience a great deal of dissatisfaction with their decision to place the baby for adoption
    c. experience considerable satisfaction with their decision and blame any dissatisfaction they feel on the infant's biological father "abandonment"
    d. experience considerable satisfaction with their decision in the immediate future, but face devastating self-blame in their middle-age years

27. Which of these is not a type of adoption discussed in the text?
    a. open adoption
    b. closed adoption
    c. semi-open adoption
    d. open-closed-open adoption

28. African-Americans comprise 12 percent of the population, and _____ percent of children awaiting adoption are black.
    a. 8
    b. 10
    c. 40
    d. 62

29. Currently about _____ of current adoptions are interracial.
    a. .13 percent
    b. .60 percent
    c. 2.0 percent
    d. 8.0 percent

30. Regarding the adoption of older children and or disabled children, disruption and dissolution rates rise with:
    a. age of the child at the time of adoption
    b. age of adoptive parents at adoption time
    c. adoption into Latino families
    d. adoptive parents having their own birth-children in the same household

The following are sample short-answer essay questions--questions of the type you may be asked if your instructor uses questions like these. Even if your instructor does not use questions like these, you can help organize and consolidate your learning if you can answer these questions in a <u>well-organized</u> and <u>complete</u> manner. Do not think that brief essays are easier than lengthy essays. They can be more challenging because you are required to be brief yet complete. And after you have answered these, try making up some similar questions and then answering them. <u>Suggestion:</u> The Study Questions at the end of each text chapter make very good short-answer essay questions.

1. Summarize briefly but completely research results about religion and fertility.

2. How do children affect marital happiness?

3. What do research results indicate about the adoption of older children and disabled children?

> **Your Opinion, Please:** Surrogate motherhood is now possible and in fact occurs. But some ethical problems remain. For example, there is the race issue. Given the history of racial exploitation in this and other countries, is it appropriate for a woman of one race to carry and bear a child for a man of another race? Why does it matter? Or doesn't it? Going further, what if the mother and father of a fertilized egg are citizens of the United States but the "pure" surrogate is a citizen of another country? What should be the citizenship of the infant? What does the law say?

Essay

The following are sample essay questions-- questions of the type you may be asked if your instructor uses essay questions. Even if your instructor does not use essay questions, you can help organize and consolidate your learning if you can answer these questions in a <u>well-organized</u> and <u>complete</u> manner. Usually, the third essay question is the most challenging.

1. **Describe** the fertility trends in the United States and, at the end of your essay, **summarize** these trends in a well-worded paragraph.

2. Explore parents' costs and benefits of having children, making sure to include in your answer social pressures as one factor that should be considered.

3. Discuss the issues that make transracial adoption an issue in our society. In your essay, do not neglect the complicating factor of class differences.

Answers

Chapter Summary

a **pronatalist** bias
a period of **structural** antinatalism
made by **default**
in **vitro** fertilization
use of a **surrogate** mother

Completion

1. structural
   antinatalism
2. pronatalist bias
3. total fertility
   rate
4. subfecundity
5. artificial
   insemination
6. abortion
7. opportunity costs
8. in vitro
   fertilization

9. programmatic
   postponers
10. paradoxical
    pregnancy
11. in vitro
    fertilization
12. involuntary
    infertility
13. adoption
14. fecundity
15. semi-open adoptions

True-False

1. T
2. T
3. T
4. T
5. F
6. F
7. T
8. F
9. T
10. T

11. F
12. F
13. T
14. F
15. T
16. F
17. T
18. F
19. F
20. T

Multiple Choice

1.	c	16.	d
2.	a	17.	a
3.	c	18.	d
4.	a	19.	a
5.	c	20.	d
6.	d	21.	a
7.	d	22.	a
8.	d	23.	b
9.	d	24.	a
10.	c	25.	c
11.	d	26.	b
12.	d	27.	d
13.	d	28.	c
14.	c	29.	d
15.	a	30.	a

CHAPTER 11
Parents and Children Over the Life Course

Chapter Overview

This chapter examines some aspects of parenting.
After a general discussion of parenting, the
authors examine the diversity of parents these
days and then examine parenting in five social
classes. The chapter next looks at racial/ethnic
diversity of parenting. After exploring the five
stages of parenting, the authors explore the
parenting process and close the chapter with
consideration of parenthood as a process that
continues throughout the life cycle.

Chapter Summary

Raising children is both exciting and frustrating.
Society-wide conditions influence the
relationship, and these factors can place
extraordinary emotional and financial strains on
parents. Formerly, children were expected to
become more help and less trouble as they grew
older; today, many parents can expect just the
opposite. Most of what children need costs more
as they grow: clothes, transportation, leisure
activities, and schooling. And as children grow
older, they may adopt values their parents have
trouble accepting. Caught between parenting their
own children and themselves being the adult child
of aging parents, some middle-aged parents--the
_____ generation--may find many role
conflicts inherent in their situation.
Of course, the problems of gay parents, lesbian
parents, and never-married mothers present some
satisfactions and problems perhaps unique to such
parents, just as there are differences among
poverty-level parents, blue-collar parents, lower
middle-class parents, upper middle-class parents,
and upper-class parents. Parenting styles also
vary: those discussed in the text include the
parent as martyr, as pal, as police officer, as
teacher-counselor, and as _____, this
last of which is seen as a "bi-directive" or

_____ perspective, which regards the influence between parent and child as mutual and reciprocal.

The text discusses the transition to parenthood and the various stages of parenthood: the transition itself, parents with babies, with preschoolers, with school-age children, with adolescents. Parenting and stepparenting can be difficult today. For example, work roles and parent roles often conflict. And middle-aged parents, especially mothers, may be sandwiched between dependent children on one hand, and increasingly dependent, aged parents on the other.

Not only mothers' but fathers' roles can be difficult, especially in a society like ours where attitudes have changed so rapidly and where there is no consensus about how to raise children and how mothers and fathers should parent. In addition, both children's and parents' needs change over the course of life. One thing that does not change as children mature is their need for supportive communication from a parent. Grandparents--particularly companionate and involved ones--can be helpful.

To have better relationships with their children, parents need to recognize their own needs and to avoid feeling unnecessary guilt; to accept help from others (friends and the community at large as well as professional caregivers); and finally, to try to build and maintain flexible, intimate relationships using techniques suggested in this chapter, along with those suggested in Chapter 8.

Key Terms

You should be able to explain the concepts listed below. In your explanation, try to avoid using the concept you are explaining. You should be able to give several examples of each concept and to explain why each example is an example.

"sandwich" generation   interactive perspective
   (p. 374)   (p. 388)
shared parenting   permissiveness
   (p. 376)   (p. 396)

primary parents
  (p. 376)
"parenting alliance"
  (p. 380)
laissez-faire discipline
  (p. 386)
autocratic discipline
  (p. 387)

overpermissiveness
  (p. 396)
democratic discipline
  (p. 400)
custodial grandparent
  (p. 409)
noncustodial grandparent
  (p. 409)

Point to Ponder: In some families, it is
expected that children will
work around the house and do chores without
pay. It is felt in these families that a
parent's role is to look out for a child and
to provide spending money when a child seems
to need it. Does the child provide unpaid
labor that if done by an unrelated adult would
be disallowed by law? If so, what justifies
this? In the same way, what reasons
justify hitting or withholding a meal from
a child if the child's behavior does not meet
with parental approval? Put differently, if
society disapproves of these economic and
disciplinary activities with adult recipients,
how--specifically--does one acount for the high
approval rates for the use of these activities
toward children in families?

Completion

Complete the following sentences by selecting the
correct alternative from the terms listed above.
Some may be used more than once. Some may not be
used at all. Filling in a blank may require more
than one word.

1. Middle-aged parents who both have their own
    children and who are themselves--even though
    middle-aged--still the now-adult children of
    their older parents, often encounter many role

conflicts stemming from their situation.
These middle-aged parents are the _____
_____.

2. In the _____
   all of the family members who are involved
   have some say in the discussion and decision
   making, and it is thought that this is more
   effective than the other two forms of
   discipline discussed in this chapter.

3. In _____
   the entire power of determining rules and
   limits is placed in the parents' hands.

4. Parents can inadvertently (or sometimes
   intentionally) support and/or undermine one
   another--or give mixed responses. The fact
   that this relationship between a mother and a
   father exists and has parenting consequences
   is called the _____.

5. _____ involves
   parents letting children set their own goals,
   rules, and limits, with little or no guidance
   from parents.

6. Viewing the parent-child relationship as bi-
   directional and of mutual influence is known
   as the _____ perspective.

7. Fathers who are intimately, actively involved
   with their children, are responsible and care
   for their children day by day, are involved in
   what the text calls _____
   _____.

8. Insofar as relationships with grandchildren
   are concerned, the older parent of the
   custodial parent is termed the _____
   grandparent.

_____
Key Research Studies

LeMasters: five parenting styles
Ehrensaft: sharing parenthood
Rossi: the transition to parenthood
Cherlin and Furstenburg: grandparents

____ 1. The concept of childhood as different from adulthood did not emerge until about the seventeenth century.

____ 2. In our society the parenting role (or stepparenting role typically conflicts with the working role, and employers place work demands first.

____ 3. Today's parents raise their children in a pluralistic society, characterized by relatively few views about ways to raise children.

____ 4. The "sandwich" generation refers to those grandparents whose roles are restricted to basic but necessary custodial roles such as getting lunch ready for children to take to school.

____ 5. Blue-collar parents are an emerging minority group.

____ 6. "The parent as martyr" refers to a parent who has actually laid down his/her life for the child, dying instead of the child, as sometimes happens in relatively rare medical settings or, less frequently, in complex accidents.

____ 7. "Autocratic discipline" is associated with the police officer style of parenting.

____ 8. Blue-collar parents tend to empathize with their child's being creative, happy, and independent.

____ 9. The children of upper-class parents are not necessarily guaranteed a comfortable level of living "no matter what."

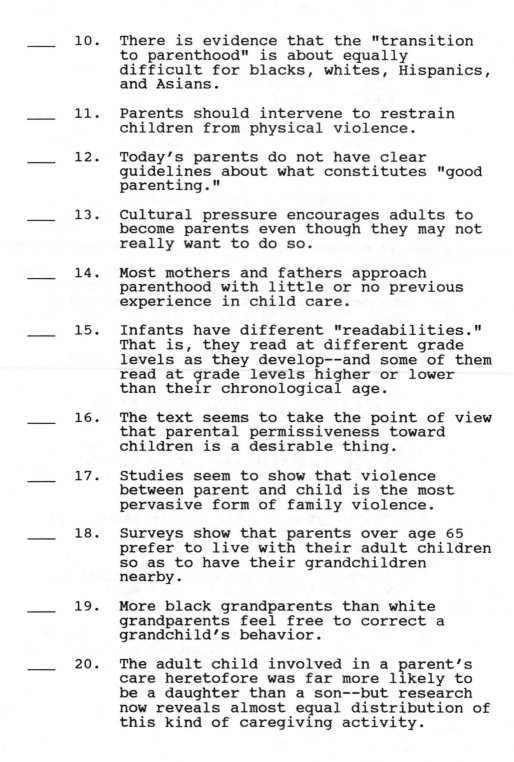

____ 10. There is evidence that the "transition to parenthood" is about equally difficult for blacks, whites, Hispanics, and Asians.

____ 11. Parents should intervene to restrain children from physical violence.

____ 12. Today's parents do not have clear guidelines about what constitutes "good parenting."

____ 13. Cultural pressure encourages adults to become parents even though they may not really want to do so.

____ 14. Most mothers and fathers approach parenthood with little or no previous experience in child care.

____ 15. Infants have different "readabilities." That is, they read at different grade levels as they develop--and some of them read at grade levels higher or lower than their chronological age.

____ 16. The text seems to take the point of view that parental permissiveness toward children is a desirable thing.

____ 17. Studies seem to show that violence between parent and child is the most pervasive form of family violence.

____ 18. Surveys show that parents over age 65 prefer to live with their adult children so as to have their grandchildren nearby.

____ 19. More black grandparents than white grandparents feel free to correct a grandchild's behavior.

____ 20. The adult child involved in a parent's care heretofore was far more likely to be a daughter than a son--but research now reveals almost equal distribution of this kind of caregiving activity.

1. Parents today are _____ to have to cope with
    serious childhood illnesses.
    a. more likely than before
    b. much more likely than before
    c. as likely today as in previous years
    d. less likely

2. The text cites a recent book's report that
    "The average employed person is now on the job
    an additional _____ hours [a year than s/he
    was two decades ago]..."
    a. 76
    b. 163
    c. 236
    d. 390

3. For a variety of reasons, parents today are
    probably judged by _____ standards than they
    were in the past.
    a. more uniform
    b. higher
    c. lower
    d. much lower

4. The "enduring image of motherhood" includes
    the idea that:
    a. a woman's identity is tenuous and trivial
        without motherhood
    b. the image of women is inextricably mingled
        with caregiving images from major
        religions
    c. significant others--including mothers--are
        indelible parts of that aspect of
        personality called the Generalized Other
    d. traditional images of mothers are
        perpetuated by the pervasive mass media
        such as television series and films

5. Studies indicate that when we control for class, the child-rearing style of African-American parents:
   a. compares to that of whites
   b. involves much harsher physical punishment when children misbehave
   c. allows children more unsupervised roaming through the neighborhood than white parents allow.
   d. is more detached than the parenting style of white parents

6. E.E. LeMasters listed how many parenting styles?
   a. four
   b. five
   c. six
   d. seven

7. Which of these is NOT one of the parenting styles listed by LeMasters?
   a. the pal
   b. the police officer
   c. the teacher-counselor
   d. the servant

8. When LeMasters and DeFrain speak of "value stretch," they refer to which of these?
   a. compromising one's values
   b. valuing pushing oneself to the limit in terms of educational/occupational achievement
   c. getting the most out of every dollar spent in the household budget
   d. a difference in the parenting behaviors of blue-collar men and their wives

9. According to the text, the accomplishments of children of _____ parents will probably never surpass those of their parents, grandparents, and great-grandparents.
   a. upper-class
   b. upper middle-class
   c. lower middle-class
   d. working-class

10. According to Gelles and Strauss, _____ can help reduce violence in families, especially between _____.
   a. day care; grandparent and now-adult child
   b. day care; parent and preschooler
   c. well-trained police; husband and wife
   d. timeout; siblings

11. Cherlin and Furstenberg found that about ____ of their sample of grandparents said they saw their grandchildren less often than every two or three months.
   a. 77 percent
   b. 54 percent
   c. 31 percent
   d. 20 percent

12. As used in the text, the term "primary parents" refers to which of these?
   a. parents who are only beginning to learn basic parenting skills
   b. parents whose children are in primary school
   c. persons who are not the biological parents but who provide basic caregiving for one or more children
   d. couples in which both parents were equally involved in "mothering" behavior

13. A study of fathers in our society found that ____ thought they **should** share child rearing equally with mothers, but ____ **actually** did so.
   a. 55 percent; 68 percent
   b. 55 percent; 21 percent
   c. 74 percent; 13 percent
   d. 80 percent; 40 percent

14. According to the text, the concept "new fathers" is associated with which of these?
   a. first-time parents
   b. conversion-oriented parents
   c. 50-50 parents
   d. newly divorced fathers

15. Parents vary:
    a. in marital status but not in sexual orientation
    b. in marital status and in sexual orientation
    c. neither in marital status nor in sexual orientation
    d. none of the above

16. A study of 242 adult children and 66 elderly parents found that the adult children tended to let parents live independently as long as:
    a. they saw each other once in a while
    b. they saw each other at least four times a month, or, approximately weekly
    c. it was safe
    d. the parents were happy with the arrangements

17. According to the text, which of the following has had the effect of leaving many more grandmothers to assume the responsibility of child rearing?
    a. the spread of AIDS and use of crack cocaine
    b. legalization of spousal separation without specification of child support and/or child care
    c. greater use of marriage dissolution and less use of divorce
    d. increased commitment to education and career by both men and women

18. Parents who loan their children large amounts of money should consider the desirability of:
    a. giving goods, merchandise, or actual property rather than cash or negotiable instruments
    b. making the loan contingent on the promise of supportive care when the parent is old and frail
    c. formalizing such loans in writing and even with the consultation of an attorney
    d. informing about the loan persons who are not family members, in order to avoid confusion should the loan-making parent(s) die intestate (that is, having no legal will)

19. In our society, parenting ends when one's child reaches age:
    a. 16-17
    b. 18-20
    c. 21-25
    d. none of these

20. In a long-term study of child development in Hawaii, the researchers spoke of a "self-righting tendency," referring to the tendency of some children to:
    a. somehow have the prerequisites to emerge into adulthood in good shape
    b. choose political and social--or traditional--options when given choices
    c. make choices that result in minimal contact with representatives of formal authorities--whether school, work organizations, or law enforcement
    d. have an exaggerated idea of their own importance in the general scheme of things, alienating many people in the process

21. Which of these is a program stressing guidelines for no-win intimacy to the parent/child relationship?
    a. LOVIT
    b. PCLOVE
    c. PET
    d. AARP

22. A program offered in many communities to assist in effective parenting is:
    a. STEP
    b. MORE
    c. NOVA
    d. MARIGOLD

23. Looking for opportunities to show the child a new picture of himself or herself, letting children overhear you say something positive about them, and being a storehouse for your child's special moments are all listed by the text as ways to:
   a. make children assertive and effective
   b. free children from playing roles
   c. increase probability that children will comply with reasonable community expectations
   d. enhance the self-image of children with diagnosed behavior problems

24. The ____ undercuts the child's development of his/her own competency; in contrast, the ____ does such things as assigns responsibility and sends positive messages about the child's competence and motivation.
   a. peer group; parent
   b. medical model; social-psychological model
   c. ritualistic teacher; effective teacher
   d. over-involved parent; consulting parent

25. According to the text, blacks are likelier than whites to:
   a. use psychological punishment for children's misbehavior
   b. use food as a reward for good behavior
   c. become grandparents at an earlier age
   d. rely on physicians rather than on books for suggestions about adequate parenting

26. "Overpermissiveness" allows undesirable ____ as ____.
   a. feelings; motivations for actions
   b. peers; role models
   c. acts; expressions of feelings
   d. role models; sources of socialization

27. According to the text, today's parents are more likely to be:
    a. remaining in the same social class as their parents
    b. experiencing social mobility, and thus coming into conflict with their own parents about the proper socialization of their children
    c. caring for fragile infants
    d. caring for parents who die of sudden, catastrophic illness

28. According to the text, today's well-educated and well-read parents:
    a. have too many resources for their own children's well-being
    b. are so fearful of possibilities that they often become too protective
    c. may overreact to normal behavior
    d. none of these

29. According to the text, parents within the same racial/ethnic group do not necessarily agree on the best approach to take to racial issues. About _____ of African-American parents do not attempt any explicit racial socialization.
    a. one-eighth
    b. one-fifth
    c. one-half
    d. one-third

30. "Zero-parent family" refers to a family in which:
    a. a parent is physically present but is absent in any meaningful caregiving sense
    b. a biological parent is not present in the home as a caregiver
    c. both parents have established a "paper trail" in the criminal justice system, effectively nullifying their capability as an economically productive role model
    d. the child remains with a biological parent who is not biologically the parent and who has not adopted the child in the usual way

Short-Answer Essay Questions

The following are sample short-answer essay
questions--questions of the type you may be asked
if your instructor uses questions like these.
Even if your instructor does not use questions
like these, you can help organize and consolidate
your learning if you can answer these questions in
a well-organized and complete manner. Do not
think that brief essays are easier than lengthy
essays. They can be more challenging because you
are required to be brief yet complete. And after
you have answered these, trying making up some
similar questions and then answering them.
Suggestion: The Study Questions at the end of
each text chapter make very good short-answer
essay questions.

1.   Is "parent" a social status? A social role?
     Both? Neither? What gives "parent" meaning?
     Biological parenthood? A social commitment?
     Or what?   Be specific--not general--in your
     response.

2.   Explain the meaning of "the parenting
     alliance."

3. How are never-married single mothers "different" from divorced, widowed, or separated mothers?

Essay

The following are sample essay questions-- questions of the type you may be asked if your instructor uses essay questions. Even if your instructor does not use essay questions, you can help organize and consolidate your learning if you can answer these questions in a _well-organized_ and _complete_ manner. Usually, the third essay question is the most challenging.

1. Explain the ways in which parenting behavior differs or varies in the various social classes.

2. Of course, everyone is in many ways individual and unique. Nevertheless, couples and individuals can expect to go through often-observed stages as they raise their children. _List_ _and_ _explain_ the stages set forth in the text.

3. How do the problems of parents with preschoolers differ from the problems of parents with young adult children? In your answer, go beyond simply saying that there are age differences. What **other** differences are there? If you can, cite specific research results to make your answer more persuasive.

Answers

Chapter Summary

the **"sandwich"** generation
and as **athletic coach**
"bi-directive" or **interactive** perspective

Completion

1.	"sandwich" generation	5.	laissez-faire
2.	democratic form of		discipline
	discipline	6.	interactive
3.	autocratic discipline	7.	shared parenting
4.	parenting alliance	8.	custodial
			grandparents

True-False

1.	T	11.	T
2.	T	12.	T
3.	F	13.	T
4.	F	14.	T
5.	F	15.	F
6.	F	16.	T
7.	T	17.	F
8.	F	18.	F
9.	F	19.	T
10.	F	20.	F

Multiple Choice

1.	d	16.	c
2.	b	17.	a
3.	b	18.	c
4.	a	19.	d
5.	a	20.	a
6.	b	21.	c
7.	d	22.	a
8.	d	23.	b
9.	a	24.	d
10.	d	25.	c
11.	d	26.	c
12.	d	27.	c
13.	c	28.	c
14.	c	29.	d
15.	b	30.	b

CHAPTER 12
Work and Family

Chapter Overview

This chapter explores traditional employment
patterns that have characterized our society. The
emphasis, of course, is on how these patterns have
affected marriage and the family. The chapter
then examines emerging patterns of employment and
the interrelationships of work and family life for
both men and women.

Chapter Summary

The labor force is a social invention.
Traditionally, marriage has been different for men
and for women. The husband's job has been as
breadwinner, the wife's as homemaker. These roles
are changing as more and more women enter the work
force. Women still remain very segregated
occupationally and earn lower incomes than men,
on the average.
    The text distinguished between two-earner and
two-career marriages. In the latter, wives and
husbands both earn high wages and work for
intrinsic rewards. Even in such marriages, the
husband's career usually has priority.
Responsibility for housework falls largely on
wives. Many wives would prefer shared roles, and
negotiation and tension over this issue cast a
shadow on many marriages. An incomplete
transition to equality at work and at home affects
family life profoundly.
    The text emphasized that both cultural
expectations and public policy affect people's
options. As individuals come to realize this,
there will be pressure on public officials to meet
the needs of working families by providing
supportive policies, including perhaps not only
the _____ track, but parental leave, child
care, and _____.
    Paid work is not usually structured to allow
time for household responsibilities. Women,
rather than men, continue to adjust their time to

accomplish both paid and unpaid work. To be
successful, _____ marriages will require
social-policy support and workplace flexibility.
But there are some things the couple themselves
can keep in mind that will facilitate their
management of a working-couple family.
Recognition of both positive and negative feelings
and open communication between partners can help
working couples cope with an imperfect social
world.

Household work and child care are pressure
points as women enter the labor force and the two-
earner marriage becomes the norm. To make it
work, either the structure of work must change,
social policy must support working families, or
women and men must change their household role
patterns--and very probably all three.

_____

Key Terms

You should be able to explain the concepts listed
below. In your explanation, try to avoid using
the concept you are explaining. You should be
able to give several examples of each concept and
to explain why each example is an example.

"good provider" role
    (p. 418)
co-provider couple
    (p. 418)
main/secondary
    provider couple
    (p. 418)
househusband
    (p. 420)
two-person single
    career
    (p. 422)
occupational
    segregation
    (p. 425)
pink-collar jobs
    (p. 425)
trailing spouse
    (p. 446)

two-earner marriage
    (p. 429)
"second shift"
    (p. 434)
gender strategy
    (p. 457)
two-career marriage
    (p. 429)
"mommy track"
    (p. 452)
family leave
    (p. 451)
flextime
    (p. 451)
commuter marriage
    (p. 446)
ambivalent provider
    couple
    (p. 418)

Point to Ponder: Sometimes couples marry
                    with the expectation that
the husband will work full-time at a job he
enjoys and that the wife will work part-time
at a job she enjoys because she wants to bring
some additional money into the household to
make life a little more comfortable. But to
their surprise, they find that to make ends
meet, both may have to settle for full-time
jobs that they do NOT particularly enjoy.
Since this was not their agreement or expectation when they married, each partner may be
deeply dissatisfied with the situation.   Why
might their economic situation come as a
surprise to them?

Completion

Complete the following sentences by selecting the
correct alternative from the terms listed above.
Some may be used more than once. Some may not be
used at all. Filling in the blanks may require
more than one word.

1. The notion that the husband or "man of the
    household" would provide all of the economic
    income was known as the _____.

2. A _____ is a man who stays at
    home to care for the house and family while
    his wife works.

3. Fred is expected to work 42 hours each week.
    But his office manager leaves it to him to
    decide how many hours and what days he will
    work to accumulate those 42 hours. This is
    the concept of _____.

4. Intentionally created part-time jobs that are
    not particularly demanding are part of the
    _____.

5. The _____ refers to the time spent by two-earner couples in negotiating and fulfilling obligations at home, together, once employment time requirements have been met.

6. A _____ is one in which the spouses live and work in geographically distant places but do spend time with each other occasionally--perhaps on weekends or at other periodic intervals.

7. An _____ means that a wife is providing responsibilities that are not clearly acknowledged.

8. In devising a _____, a person considers her/his goals and decides upon a course of action based on the expectations and norms of the culture to achieve those goals.

9. The concept of _____ means that workers are allowed to take periods of time off from work--sometimes paid--to tend to family matters such as pregnancy, a parent's terminal illness, or a period of intensive child care.

10. A(n) _____ is one who relocates to accommodate the other's (but not one's own) career.

Key Research Studies

You should be familiar with the main question being investigated and the research results for the following studies:

Gore and Mangione: stresses of the homemaker role (in Anglo and Mexican-American cultures, for example)
U.S. Bureau of the Census: statistics on work and occupation, throughout the chapter
Lewin: day care

Conflict Theory: Why Wives Do the Housework
  (Box 12.1)

True-False

____ 1. The "good-provider role" is a relatively
          new one in American society.

____ 2. Today, although occupational and sex
          discrimination in employment persist, the
          trend is away from distinguishing work
          based on sex.

____ 3. One of the rewards of the good-provider
          role for husbands has been reinforcement
          of patriarchal power in the family.

____ 4. People who want new family patterns
          should be aware of cultural expectations
          and their own feelings about such
          choices.

____ 5. In the main/secondary provider couple,
          providing is the man's primary
          responsibility, and the home is the
          woman's primary responsibility.

____ 6. Couples who have been married for a
          shorter time seem to have more
          difficulties with "commuter marriages."

____ 7. If wives are employed, it means that
          their husbands have transferred
          psychological responsibility for
          providing for the family (or it means
          that such responsibility is shared).

____ 8. Recent research indicates that co-
          provider couples are about 30 percent of
          all unions.

____ 9. About six percent of husbands are
          househusbands or stay-at-home dads.

_____ 10. Early research supported the notion that working mothers were not detrimental and might even be advantageous to their children's development. This continues to be the prevailing view.

_____ 11. According to the text, "currently more than one-third of married women are full-time homemakers."

_____ 12. One study of 119 major corporations found that about 40 percent of the corporations believed that "no time" was the appropriate amount of time for a male employee to take off at the birth of his child.

_____ 13. In the United States today only a small minority of states have "community property" laws in which all of the couple's property is considered to be owned equally and in common.

_____ 14. In the United States today, the most acceptable pattern is for the wife to stay home if any spouse does.

_____ 15. "Unpaid leave" is a child-care program that pays parents to allow their children to take part in summer enrichment programs occurring outside the parental home.

_____ 16. Children who stay at day-care centers have a significantly higher level of illness and injury than children who are left at home with a friend, neighbor, or relative.

_____ 17. Mexican culture places more importance on the family than on work for both sexes.

_____ 18. REACT is an organization that is related "reasonable events (in the life of) a Christian team of partners."

_____ 19. White women are now beginning to catch up with black women in terms of percent of participation in the labor force.

_____ 20. Working women tend to get more household help from their husbands when the women's working is perceived to be necessary for the economic welfare of the household.

Multiple Choice

1. The trend is clearly away from distinguishing work based on:
   a. skill levels
   b. pay or wage levels
   c. sex or gender
   d. seniority

2. A serious "cost" of the good-provider role was that:
   a. the statuses attached to the role were confused and overlapping
   b. gender identification depended only on this one role
   c. gender identification became almost impossible, though sex identification was untouched
   d. the fines and court judgments were too expensive to continue with the role

3. When a wife spends much time and energy helping her husband in his career, this is what is meant by a(n):
   a. facilitating couple
   b. two-person single career
   c. sideline wife
   d. "organizer"

4. Society encourages men to give primacy to their _____ and let _____ come second.
   a. wife; children
   b. children; wife
   c. work; fictive kin
   d. work; family

5. The last women to move into employment outside the home have been:
   a. women recently widowed
   b. divorced women
   c. young women
   d. mothers of young children

6. The pronounced tendency for men and women to be employed in different types of jobs is called:
   a. job prejudice
   b. ability sifting
   c. mixed-ability job statuses
   d. occupational segregation

7. A concept associated with househusbands is:
   a. dysfunctional deviance
   b. latent avoidance
   c. role reversal
   d. gender exploitation

8. In a study of 317 mothers of infants up to three months after their infants' births, researchers found that ____ of mothers felt that they would rather stay home with the baby than seek outside employment.
   a. 20-25 percent
   b. 40-45 percent
   c. 50-55 percent
   d. 70-75 percent

9. The "second shift" is connect in your text with:
   a. two-job husbands
   b. two-job wives
   c. the leisure gap
   d. beginning to have one's children later rather than earlier

10. About ____ of workers have flexible schedules, allowing them to choose their work hours--within some limits.
   a. 12 percent
   b. 22 percent
   c. 35 percent
   d. 48 percent

11. In the United States, historically,
    _____ have been more likely to work for wages
    than have _____.
    a. wives of upper white-collar men; wives of
       lower white-collar men
    b. wives of lower white-collar men; wives of
       upper white-collar men
    c. well-educated women; less well-educated
       women
    d. African-American women; white women

12. According to the text, women-dominated
    professions tend to be:
    a. professions with seasonal work
    b. occupations requiring lower rather than
       higher amounts of decision making
    c. service or support professions
    d. high-tech, "clean" occupations

13. The vast majority of single women are
    employed:
    a. as a temporary stage viewed as a prelude
       to marriage--in Japan, termed "diving-
       board girls"
    b. in order to supplement the economic
       requirements of their parental household
    c. in order to support themselves
    d. as a distraction to boredom while they
       wait for "the standard package" of
       marriage and family life

14. In the United States, _____ of husbands do as
    much housework as their wives.
    a. 75-80 percent
    b. nearly 70 percent
    c. about half
    d. one-tenth or fewer

15. Hochschild uses the concept "stalled
    revolution" to refer to which of these?
    a. equalizing the gender difference in
       income for similar jobs
    b. employers providing child care at the
       work site
    c. amount of government concern about
       all employees' rights for domestic leave,
       regardless of sex
    d. increases in husbands' time spent doing
       household work

16. According to the text, a "reinforcing cycle" is associated with men's and women's:
    a. work
    b. involvement in child rearing
    c. boundary maintenance in gender roles
    d. responsibilities for children's formal education experiences

17. According to the text, which of the following is sometimes or often an issue affecting a two-career family?
    a. needs for independence and autonomy
    b. geography
    c. medical needs
    d. none of these

18. The concept of a "trailing spouse" refers to a spouse who:
    a. relocates to meet the requirements of the other's career
    b. is the last to finish formal educational requirements for job advancement
    c. faces insecurities reducing him/her to using professional agencies to check up on the other's activities
    d. advances at the same rate occupationally but at a slower rate socially than the other spouse

19. As an alternative to the "mommy track," one analyst has suggested the "life course solution," which would require _____ the pattern of education, employment, and retirement.
    a. increased measurement or quantification of
    b. solidifying
    c. rethinking
    d. eliminating

20. This chapter concludes by stating that the adjustment of individual family members, as well as harmonious family relationships, requires _____ the very different and often conflicting needs of family members.
    a. balance between
    b. legitimation of
    c. formal legal recognition of
    d. elimination of

> **Your Opinion, Please:** Ted, an executive for
> an electronics company
> in Ohio, is married to Carol, who is not
> employed, as the couple agreed before they
> married. Ted's company decided to relocate his
> job to California. Carol, though not thrilled
> with the idea, moved with him to California to
> where she opened what turned out to be a
> highly successful mail-order business from
> their home. Ted's company then transferred his
> job back to Ohio. Carol saw no compelling
> reason to relocate, asking Ted to remain in
> California, since her income was now double
> his, and refused to relocate to Ohio. Ted
> returned to Ohio and sued his wife for divorce
> on grounds of desertion. The court awarded him
> a divorce on these grounds. What is your
> opinion about this case? Do you think Ted
> should have remained in California? Do you
> think Carol should have returned to Cleveland?
> Do you think Ted's court case was appropriate
> or do you see a more desirable alternative?

Short-Answer Essay Questions

The following are sample short-answer essay
questions--questions of the type you may be asked
if your instructor uses questions like these.
Even if your instructor does not use questions
like these, you can help organize and consolidate
your learning if you can answer these questions in
a <u>well-organized</u> and <u>complete</u> manner. Do not
think that brief essays are easier than lengthy
essays. They may be more challenging because you
are required to be both brief yet complete. And
after you have answered these, try making up some
similar questions and then answering them.
Suggestion: The Study Questions at the end of
each text chapter are very good short-answer essay
questions.

1. For husbands, what are the rewards and costs
   of the provider role?

2. What are the **differences** in the reasons single women and married women are in the labor force?

3. What is "the life course solution" to day care? Explain briefly but completely.

Essay

The following are sample essay questions-- questions of the type you may be asked if your instructor uses essay questions. Even if your instructor does not use essay questions, you can help organize and consolidate your learning if you can answer these questions in a well-organized and complete manner. Usually, the third essay question is the most challenging.

1. Describe--and then compare and contrast--the work realities of husbands and of wives.

2. When couples change from being one-earner couples to being two-earner couples, what are the positive and negative consequences for them?

3. Cecilia and Andre are both fully employed professionals who intend to marry. They both intend to continue their professional careers. What do sociologists who study families know

that would probably be useful for Cecilia and Andre? Be specific, be complete, and cite specific research findings to increase your answer's credibility.

Answers

Chapter Summary

the **mommy** track
child care, and **flextime**
two-earner marriages

Completion

1. good-provider
2. househusband
3. flextime
4. "mommy track"
5. "second shift"
6. commuter marriage
7. ambivalent provider couple
8. gender strategy
9. family leave
10. trailing spouse

True-False

1.	F	11.	T
2.	T	12.	T
3.	T	13.	T
4.	T	14.	T
5.	T	15.	F
6.	T	16.	F
7.	F	17.	T
8.	F	18.	F
9.	F	19.	T
10.	T	20.	T

Multiple Choice

1.	c	11.	d
2.	b	12.	c
3.	b	13.	c
4.	d	14.	d
5.	d	15.	d
6.	d	16.	a
7.	c	17.	b
8.	d	18.	a
9.	c	19.	c
10.	a	20.	a

CHAPTER 13
Managing Financial Resources

Chapter Overview

Chapter 13 discusses how families provide for their basic needs--shelter, food, clothing, and health care. The chapter focuses on earning and spending, budgets and credit, as well as family housing and health care.

Most American families experience a series of financial ups and downs. Much is experienced as personal troubles. But there is a need to realize that individual, couple, and family economic difficulties are affected by government policies. The chapter explores how individuals, couples, and families can take an active role in influencing public policies about the allocation of social resources.

Chapter Summary

The economic pinch that our society has been experiencing in the past few decades affects family life in significant ways. We saw in earlier chapters that more wives are going to work, partly to increase the family income. This chapter looks at things families can do to spend most efficiently the money they earn.

Budgets help families to manage expenditures wisely. <u>Planning budgets</u> is not all that's needed, though; <u>daily</u> <u>and</u> <u>monthly</u> _____ should be kept as well. <u>Buying wisely</u> is also important, for _____ <u>consumers</u> can get the most for their money. _____ expenditures should be budgeted, just like any other spending.

Two critical family needs are housing and health care. Costs of both have soared in recent years, posing real challenges to families. Families who make a serious effort to **not** make consumer decisions by _____ are best able to meet those challenges.

Many of the personal or private troubles families have with finances, housing, food costs, or health care are really _____ issues, and families can join together to change them.

Key Terms

You should be able to explain the concepts listed below. In your explanation, try to avoid using the concept you are explaining. You should be able to give several examples of each concept and to explain <u>why</u> each example is an example.

budgeting
   (p. 475)
discretionary income
   (p. 476)
installment financing
   (p. 479)
consolidation loan
   (p. 480)
neo-local residence
   (p. 481)

family policy
   (p. 488)
medicaid
   (p. 492)
medicare)
   (p. 492)
social security
   (p. 494)

Point to Ponder: For the most part, American norms specify neo-locality for newlyweds--the married couple is expected to live in their own residence, apart from parents. To what extent do you think that the high cost of housing--and other living expenses-- <u>has</u> made it difficult for couples to form relationships that conform to the residence norms? To what extent does the cost of housing affect the fertility rate? Or is the cost of housing not really a problem? What <u>does</u> the <u>text</u> suggest about this issue?

Completion

Complete the following sentences by selecting the correct alternative from the terms listed above. Some may be used more than once. Some may not be used at all. Filling in a blank may require more than one word.

1.   With _____, individuals and families can discover and plug leaks in their expenditures.

2. With _____, it is important to find out how much the item will really cost, that is, the price of the item plus the hidden cost of financing.

3. The _____ is one method of coping with unwise borrowing behavior.

4. The _____ residence pattern expects that newlywed couples leave their parents' homes and set up their own households.

5. One out of every six Americans receives _____ benefits.

_____

Key Research Studies

You should be familiar with the main question being investigated and the research results for the following studies.

U.S. Bureau of the Census: data on various topics in the general area of family economics

_____

True-False

____ 1. Real income for Americans peaked in 1973.

____ 2. The wages of blue-collar workers in the United States are unstable and declining.

____ 3. Recent economic changes have about equally hurt white and minority families.

____ 4. Most American families achieve long periods of financial stability. A sizable minority of them experience what the text calls "a series of financial ups and downs."

____ 5. A child age 5 or younger is less expensive to provide for than is a child age 12 to 17.

_____ 6. Discretionary income is income that results from the use of exceptionally good judgment about purchasing essential household goods.

_____ 7. Step three in planning a budget is "estimating your income."

_____ 8. Step five in planning a budget is "prepare the budget."

_____ 9. In the context of this chapter--managing financial resources--the term "compounding" was used to refer to making one's problems more difficult by postponing them as long as possible.

_____ 10. The text states that to boost one's morale or to try to increase one's prestige is the wrong reason for borrowing.

_____ 11. The textbook for this course urges readers not to use credit counselor services because they charge fees that only increase one's debt burden.

_____ 12. A consolidation loan seems to work for most people if they change their credit-charging habits.

_____ 13. A recent federal law makes declaring bankruptcy more difficult than it used to be.

_____ 14. The declaration of bankruptcy will remain on one's credit record for ten years.

_____ 15. The U.S. government has a Fair Housing law that forbids low-income renters to be charged rent greater than 33 percent of their total income.

_____ 16. Cost-of-living increases have not yet been built into Social Security payments.

Multiple Choice

1. Real income for American families peaked in:
   a. 1955
   b. 1965
   c. 1973
   d. 1981

2. The 90s have been characterized by a longer-lasting:
   a. balance of trade stability
   b. period of economic growth
   c. synchronous purchasing cycle
   d. recession

3. The proportion of the American population living in poverty increased in the 1980s to a peak of _____ in _____.
   a. 3 percent; 1989
   b. 15.2 percent; 1983
   c. 7.6 percent; 1990
   d. 34 percent; 1988

4. African-American families have only ___ of the amount of assets that white families have.
   a. 10 percent
   b. 35 percent
   c. 55 percent
   d. 70 percent

5. Over the last twenty years, the proportion of black families headed by women increased, while the proportion of _____ fell.
   a. white families headed by women
   b. single black men
   c. black men with jobs
   d. single black women

6. Our image of the poor as largely confined to inner city minority populations is:
   a. increasingly confirmed by research
   b. mistaken
   c. underestimating the extent to which this occurs
   d. a result of television and print journalism educating the public about reality

7. Which of the following is an example of those among whom poverty is now concentrated?
    a. Sandra, age 24, works as a medical secretary and struggles to pay her rent.
    b. Freddie, age 12, lives with his dad and contributes to the household income by working his paper route.
    c. Lisa, age 7, lives with her single mom.
    d. Andrew, age 78, lives with Fred, age 82, and together they struggle with the economic realities of old age.

8. About ___ of black children are poor.
    a. 76 percent
    b. 40 percent
    c. 27 percent
    d. 19 percent

9. Poor children may drop out of school to try to help their families. This is one explanation for the especially high dropout rate of:
    a. Chinese-American youth
    b. Latino youth
    c. youth who emigrated from rural to urban areas
    d. Japanese-American youth

10. The gap between very wealthy and middle-income families has:
    a. been largely stable during the last decade
    b. widened
    c. been narrowing persistently since 1975
    d. been narrowing persistently since 1982

11. At higher levels of the middle class, _____ has not kept up with inflation since 1989.
    a. the pay of college graduates
    b. the risk/benefit ratio of home ownership
    c. the value of land
    d. financial contribution to religious organizations

12. In the United States these days both spouses
    are often employed, and this is a result of:
    a. men exploiting their wives
    b. constantly expanding desires related to
        level of living
    c. economic need rather than of life-style
    d. women's desire to leave undesired
        household chores and embrace paid
        employment

13. Which of these does the text seem to recommend
    in its chapter on managing financial
    resources?
    a. avoiding credit consultation services
    b. using credit cards up to the credit limit
    c. using several credit cards, spreading
        one's charges over all of them
    d. prenuptial financial planning

14. The average American middle-income couple
    experiences a declining economic burden when
    children become self-supporting. But
    financial security may vanish again when:
    a. the grandchildren are born
    b. the parents retire
    c. self-supporting children become suddenly
        and seriously ill
    d. the parents divorce between age 50-65

15. According to the text, at retirement, income
    can drop as much as:
    a. 5-6 percent
    b. 10-12 percent
    c. 50 percent
    d. 21 percent

16. According to the text, _____ are better off,
    on the average, than young Americans.
    a. older people
    b. young Canadians
    c. young people in Great Britain
    d. none of these

17. According to the text, because older men are more apt to have living spouses, they are also more likely to be in a:
    a. better quality nursing home
    b. situation providing them with more transportation options
    c. two-income family
    d. situation where physical abuse occurs

18. Later cohorts of retiring women are likely to be better situated than in the past because of:
    a. their higher level of assertiveness
    b. their extensive labor force participation
    c. their choices to return to occupations traditionally selected by women
    d. their longer life expectancy

19. The second step in preparing a budget is:
    a. preparing a budget
    b. estimating income
    c. comparison shopping
    d. none of these

20. Which of these is a tax-deferred savings plan?
    a. AMITA
    b. L-trust
    c. SRA
    d. Chapter 10

21. Some couples with credit problems try a(n) ___ to help them cope with their financial problems.
    a. bank hyperdraft
    b. money market fund
    c. AMITA
    d. debt consolidation loan

22. A(n) _____ is a single loan that is large enough to cover all outstanding debts.
    a. letter of credit
    b. bank draft loan
    c. escrow loan
    d. consolidation loan

23. The Fair Debt Collection Practices Act
    provides protection for overextended
    consumers. Among other things, the law
    prohibits independent collection agencies
    from all of the following except:
    a. phoning before 9 a.m. or after 11 p.m.
    b. contacting your employer to verify where
        you work
    c. contacting your friends to verify where
        you live
    d. making abusive or threatening phone calls

24. Forty-seven percent of low-income renters pay
    _____ or more of their income for housing.
    a. 20 percent
    b. 45 percent
    c. 58 percent
    d. 70 percent

25. As hospital patients are discharged more
    rapidly from hospitals due to regulatory
    practices or health insurance specifications,
    it is _____ who have picked up the slack,
    providing skilled medical services that would
    be quite expensive if billed.
    a. public health nurses
    b. free clinics
    c. the friendship system rather than the
        kinship system
    d. women

Short-Answer Essay Questions

The following are sample short-answer essay
questions--questions of the type you may be asked
if your instructor uses questions like these.
Even if your instructor does not use questions
like these, you can help organize and consolidate
your learning if you can answer these questions in
a <u>well-organized</u> and <u>complete</u> manner.   Do not
think that brief essays are easier than lengthy
essays.   They may be more challenging because you
are required to be both brief yet complete.   And
after you have answered these, try making up some
similar questions and then answering them.
Suggestion: The Study Questions at the end of
each text chapter are very good short-answer essay
questions.

1.   Briefly, which Americans are increasingly
     likely to live in poverty?

2. Summarize earning and spending patterns over the course of family life in the United States.

3. Assess the potential of credit purchases as a part of the family economy.

Essay

The following are sample essay questions-- questions of the type you may be asked if your instructor uses essay questions. Even if your instructor does not use essay questions, you can help organize and consolidate your learning if you can answer these questions in a well-organized and complete manner. Usually, the third essay question is the most challenging.

1. In what ways does society's economic system affect the family's economic resources in various stages of the family life course?

2. What realities and constraints face couples as they go about providing housing and health care for themselves and for their children?

3. In what ways do problems at the level of the national economic system, government family policy, and individual family economics affect each other? In your answer, be sure to address relationships between these three areas--do not just describe each area in isolation from the others.

Answers

Chapter Summary

daily and monthly **records**   decisions by **default**
informed consumers   **public** issues
credit expenditures

Completion

1. budgeting
2. installment
     financing
3. consolidation loan

4. neo-local
5. Social Security

True-False

1.	T	9.	F
2.	T	10.	T
3.	F	11.	F
4.	F	12.	T
5.	T	13.	F
6.	F	14.	T
7.	T	15.	F
8.	T	16.	F

Multiple Choice

1.	c	14.	b
2.	d	15.	c
3.	b	16.	a
4.	a	17.	c
5.	c	18.	b
6.	b	19.	d
7.	c	20.	c
8.	b	21.	d
9.	b	22.	d
10.	b	23.	a
11.	a	24.	d
12.	c	25.	d
13.	d		

CHAPTER 14
Managing Family Stress and Crises

Chapter Overview

This chapter describes changing family situations or crises and what families can do to meet them creatively and effectively. The origin of the word "crisis" comes from the Greek language, and it means "a dangerous opportunity." A crisis, then, is a turning point that has interesting possibilities, both positive and negative. It is a time of relative instability. Crises in families can be any changes that results in a difference in what people expect from one another. Both unusual events and common events can precipitate crises. Some crises are predictable, and because they are predictable they are called "transitions" rather than full-fledged crises. The chapter views transitions as predictable crises. It also examines how families define crises and how their definitions affect what happens during a crisis. The chapter ends by looking at some ways of meeting crises creatively.

Chapter Summary

Family stress is a state of tension that arises when demands tax a family's resources. Whether financial pressures or emotional support for ill family members, family stress calls for family adjustment. Many times, though, there is no easy way to remove pressures of time, money, or responsibilities. Family coping can take a social psychological form. And constant stress can be corrosive to family members and family harmony.

Also, families can experience a family _____, which can be defined as a crucial change in the course of events, a turning point, an unstable condition in affairs--all in all, a sharper trauma to the family.

Family crises may be expected and normative, as when a baby is born or adopted, or they may be unexpected. In either case the event that causes the crisis is called the _____. These may

be of various types and have varied
characteristics. Generally, when these are
expected (normative), brief, external, defined by
concrete norms, and improving, they are less
difficult to cope with.

The predictable changes of individuals and
families--parenthood, midlife transitions,
postparenthood, retirement, and widow- and
widowerhood--are all family transitions that can
be viewed as stressors. During transitions,
spouses can expect their relationship to follow
the course of a family crisis.

A common pattern can be traced in families'
reactions to crises. The three phases are the
_____ itself, the period of _____
that follows, and the reorganization or recovery
phase after the family reaches a low point. The
eventual level of reorganization a family reaches
depends on a number of factors, including the type
of _____, the degree of stress it
imposes, whether it is accompanied by other
stressors, and the family's definition of the
crisis situation. Various models of family crisis
and reorganization seek to capture this process
analytically and to pursue through research useful
knowledge about coping with crisis.

While they have options and choices, however,
family members do not have absolute control over
their lives. Many family troubles are really the
results of public issues. The serious family
disorganization that results from unemployment or
poverty, for example, is as much a social as a
private problem. Your text has often suggested
that the tension between individualistic and
family values in our society creates personal and
family conflict. And the society-wide movement
toward equality and greater self-actualization may
actually spark family crises as wives and children
move toward independence. Moreover, many family
crises are harder to bear because communities lack
adequate resources to help families meet them.
Families must act collectively to obtain the
social resources they need for effective crisis
management in everyday living.

Meeting crises creatively means resuming
daily functioning at or above the level that
existed before the crisis. Several factors can
help: a positive outlook, spiritual values, the
presence of support groups, high self-esteem, open

and supportive communication within the family, adaptability, counseling, and the presence of a kin network.

Key Terms

You should be able to explain the concepts listed below. In your explanation, try to avoid using the concept you are explaining. You should be able to give several examples of each concept and to explain why each example is an example.

family stress
   (p. 502)
crisis
   (p. 502)
transition
   (p. 502)
stressor
   (p. 502)
stressor overload
   (p. 505)
empty nest
   (p. 507)
bereavement
   (p. 510)
period of
   disorganization
   (p. 512)

ABC-X model
   (p. 515)
double ABC-X model
   (p. 515)
pile-up
   (p. 515)
resiliency model of
   family stress, adjust-
   ment, and adaptation
   (p. 516)
strong families
   (p. 516)
weak families
   (p. 517)

Point to Ponder: Think for a moment about the activities singles take part in as they date, go out, or get together. About what percent of these activities, which eventually may lead to marriage, give any clue as to how well or how poorly a potential spouse handles crises? In most cases, crisis-handling ability is something that goes largely unobserved until after marriage or cohabitation has occurred. What effects do you think this has on sustaining such relationships? Can you think of a way to ensure that crisis-handling ability can be observed **before** marriage occurs?

Complete the following sentences by selecting the correct alternative from the terms listed above. Some may be used more than once. Some may not be used at all. Filling in a blank may require more than one word.

1. A _____ necessarily involves change.

2. Carol and Joe have a pleasant life with many pleasures and few strains. One day Joe's boss tells him that further promotions will require that Joe get a Master's of Business Administration degree. Joe's boss's announcement is an example of a(n) _____.

3. Not only did Joe enter school but his mother died, which interrupted her support of Joe's education. One of the children also needed to spend the winter in Spain for an academic exchange program. Carol lost several of her major clients at the advertising agency, through no fault of her own, and eight other similar things made life difficult for this couple. They are suffering from

    _____.

4. Carol was expecting that Joe would be required to get more education. Joe shared this expectation but hoped that it would be later rather than sooner. The fact that they expected it helps define it as a(n):

    _____.

5. Characteristically, _____ creeps up on people without their realizing it.

6. The _____ is much like the double ABC-X model, and elaborates on it.

7. "Family pile-up" (prior family hardships and strains that continue to affect family life) is part of the _____.

8. The _____
   states that the stressor event (or A)
   interacts with the family's ability to cope
   with a crisis (or B), which interacts with
   the family's definition of the event, which
   produces the crisis.

9. In coping with a crisis, it is in the
   _____ that family
   members face the decision of whether to
   express or to smother any angry feelings they
   may have.

10. The _____ is one
    transition with a high likelihood of a
    positive outcome.

11. The death of a much-loved great-grandparent
    can ordinarily be expected to be followed by
    _____.

12. Shortness of breath, frequent sighing,
    tightness in the chest, and loss of energy
    are all associated with _____.

Key Research Studies

You should be familiar with the main question
being investigated and the research results for
the following studies:

   Aldous: adjustment to retirement
   Miller and Knapp: communication with the dying

Key Theories

Hill: the ABC-X model
McCubbin and Patterson: the double ABC-X model
H. and M. McCubbin: the resiliency model of family
   stress, adjustment, and adaptation
Parsons and Lidz: death as a transition point

____ 1. A crisis is a time of relative instability.

____ 2. Common events can precipitate a crisis.

____ 3. The ABC-X model is applicable to an INDIVIDUAL, whereas the double ABC-X model is applicable to a married couple or to a cohabiting PAIR.

____ 4. The loss of potential children through miscarriage meets the text's definition of a crisis.

____ 5. The third phase of a crisis is the period of disorganization.

____ 6. Box 14.2, Alcoholism as a Family Crisis, presents alcoholism as a crisis with seven stages.

____ 7. In the United States in the 1750s--before the Revolutionary War--parents were especially distraught and bereft when a child died.

____ 8. Current research paints a picture of mothers in the empty-nest stage as being predominantly depressed and as having free-floating anxiety, a reaction that is not unusual but that is almost unvaryingly negative in its implications for mental health.

____ 9. In a study of seventy-four husbands with multiple sclerosis, most of them responded to their disease by becoming "spectators" in their own homes.

____ 10. As more women work outside the home, couples' adjustments to retirement may go somewhat more smoothly.

_____ 11. The better a couple adjusts to retirement, the less painful may be a forthcoming transition to widow- or widowerhood.

_____ 12. Widowhood is significantly less common in our society than is widowerhood.

_____ 13. Bereavement is a period of mourning.

_____ 14. Strong relationships with friends are related to high morale among widows.

Multiple Choice

1. A crisis is defined by the fact that:
   a. it has a distinctly negative outcome
   b. it has the potential for a distinctly negative outcome only
   c. it is a dramatic change from usual family relationships
   d. it is not caused by the family or by its individual members

2. In a crisis situation, there is:
   a. the prospect that things will never again be the same
   b. a period of relative instability
   c. the probability that family relationships will change for the worse
   d. a perception by the general community that the family is in some ways deviant

3. According to the text, a family stressor:
   a. occurs when the family is not ready to meet it because of overload
   b. has the potential to precipitate a crisis
   c. occurs when the family has to distinguish between two categories of events
   d. happens because outside events are always affecting the family

4. A "family pile-up" is part of:
    a. most people's life, especially during early marriage
    b. the familial culture lag model of family development
    c. the double ABC-X model of family crisis
    d. the pre-stressor syndrome

5. External stressors are:
    a. those that the family creates for itself
    b. those that are external to the individual but not to the family itself
    c. those that come to the family from the external environment
    d. those that come from the physical environment rather than from the psychological environment

6. The concept of "pile-up" refers to which of the following?
    a. experiencing multiple stressors
    b. having several children, not just two
    c. having many bills to pay and paying them all at once
    d. family reunions

7. Families who _____ suffer more as individuals and also tend to provide less support for one another.
    a. oppose instrumental relationships
    b. have never faced a real crisis before
    c. define a problem as their fault
    d. have lower levels of formal education

8. In Box 14.2, Alcoholism as a Family Crisis, the third stage is:
    a. attempt to eliminate the problem
    b. denial of the problem
    c. disorganization
    d. reorganization with sobriety

9. In Box 14.2, Alcoholism as a Family Crisis, the fifth stage is:
    a. attempts to eliminate the problem
    b. denial of the problem
    c. disorganization
    d. efforts to escape the problem

10. A vertical family is one in which:
    a. family pile-up mounts higher and higher
    b. there are attachments between family generations
    c. family members communicate verbally but not emotionally
    d. family members communicate emotionally but not verbally

11. The word "crisis" comes from the Greek word for:
    a. inflexibility
    b. fright
    c. lack of practical experience
    d. decision

12. According to the text, _____ whites are likelier to live in extended family grouping than are whites in the other social classes.
    a. lower-class
    b. pink-collar class
    c. working-class
    d. white-collar class

13. According to the text, _____ has a _____.
    a. helping; dark side
    b. crisis management; downward spiral
    c. crisis management; mind-set based on an industrial model
    d. ethnicity; negative effect on coping ability

14. According to the text, _____ more than _____ have difficulty coping with their severely retarded children.
    a. older parents; younger parents
    b. black parents; white parents
    c. remarried couples; first marriage couples
    d. fathers; mothers

15. Which of the following can help families after a crisis?
    a. counseling
    b. locating the person whose fault the crisis was
    c. counseling the person whose fault the crisis was
    D. making available a low-cost method to terminate family relationships

16. The resiliency model of family adjustment and adaptation incorporates the "Double ABC-X Model of Family Stressors and Strains," but adds the elements of family system and family typology. Family typology is of two types:
    a. strong or weak
    b. out-reaching or inner-directed
    c. resourceful or without coping skills
    d. troubled or untroubled

17. According to the text, the "typical American family" has:
    a. access to local governmental resources to which it fails to request access
    b. access to state governmental resources to which it fails to request access
    c. a low level of coping ability in areas of politics and economics
    d. a high level of stress at all times

18. According to the text, even when it is deserved, _____ is less productive than viewing the crisis primarily as a challenge.
    a. giving up
    b. thinking innovatively
    c. casting blame
    d. seeking legal remedies

19. Miller and Knapp, who have done extensive research with the dying, generally endorse which of the following approaches?
    a. a quasi-symbolic approach
    b. a quasi-functional approach
    c. a dialectical approach
    d. a symbiotic approach

20. Which of the following has been correlated with longer survival of terminal patients?
    a. close geographical proximity to kin
    b. open communication styles
    c. an extended family system
    d. close but flexible relationship with caregiving physicians

21. According to the text, women who adapted to husbands' unemployment by seeking emotional support from relatives and friends found that their husbands:
    a. felt uncomfortable with the informal nature of the arrangements
    b. insisted on formal rather than informal arrangements
    c. were comforted by this indication of micro-community support
    d. viewed this as disloyalty

22. A study of alcoholic families found that _____ can serve as effective tools to cope with family stress and crisis.
    a. substructural community support groups
    b. pan-cultural value systems
    c. rituals
    d. informal contacts with community service agencies

23. Extended families consist of parents and their children who live in the same households with other relatives, such as the parents or a brother or sister of one of the spouses. Just under _____ percent of American family households are extended-family households.
    a. 4
    b. 9
    c. 17
    d. 23

24. Black married couples are _____ than white married couples to live with extended kin.
    a. no more likely
    b. twice as likely
    c. four times as likely
    d. seven times as likely

25. While in most racial/ethnic groups the extended family is viewed as a resource in times of trouble, some clinicians argue that in _____ families, having the family find out about your distress may be the problem rather than the solution.
   a. working-class
   b. upper-class
   c. value-bonded
   d. Irish

26. These days, increasingly, caregivers themselves are:
   a. co-dependents
   b. increasingly elderly
   c. financially more able to support those dependent upon them
   d. equal partners with the government to support those dependent upon them

27. The text specifically mentions _____ as disrupting or weakening some family ties that might have been resources.
   a. divorce
   b. military relocation
   c. compliance with government regulations
   d. poor communication patterns

Your Opinion, Please: The notion that family problems ought to remain _family_ problems--sometimes called privatization--may be a mixed blessing. It may ensure family privacy, but this privatization may make coping with crises more difficult. Which would **you** rather have: more family privacy but less help from friends and neighbors in coping with crises, or less family privacy but more help from friends and neighbors in coping with crises? Do you think it is possible to have both? If you think you _can_ have both, how--precisely--is it possible to have both?

The following are sample short-answer essay questions--questions of the type you may be asked if your instructor uses questions like these. Even if your instructor does not use questions like these, you can help organize and consolidate your learning if you can answer these questions in a <u>well-organized</u> and <u>complete</u> manner. Do not think that brief essays are easier than lengthy essays. They can be more challenging because you are required to be brief yet complete. And after you have answered these, try making up some similar questions and then answering them. Suggestion: The Study Questions at the end of each text chapter make very good short-answer essay questions.

1.   What is stressor overload? Give two examples.

2.   Explain briefly but completely how the postparental period can be a stressor.

3.   Distinguish between "high self-esteem" and "having a positive outlook" as factors in meeting crises effectively.

Essay

The following are sample essay questions--
questions of the type you may be asked if your
instructor uses essay questions. Even if your
instructor does not use essay questions, you can
help organize and consolidate your learning if you
can answer these questions in a <u>well-organized</u> and
<u>complete</u> manner. Usually, the third essay
question is the most challenging.

1. Probably everybody has problems. But what is
    the distinction between having problems and
    having a crisis? Explore, making specific
    reference to material in the text.

2. What are the various stages of coming to grips
    with a crisis? And, do you think that people
    go through the same stages regardless of the
    crisis with which they are attempting to cope?
    Explore the question, but remember that you
    should support your answer with material from
    the text.

3. Compare and contrast the ABC-X model, the
    double ABC-X model, and the resiliency model
    of family stress, adjustment, and adaptation.

Answers

Chapter Summary

a family **crisis**
is called the **stressor**
the **crisis** itself
the period of **disorganization**
the type of **stressor**

Completion

1.	crisis	7.	double ABC-X model
2.	stressor	8.	ABC-X model
3.	stressor overload	9.	period of disorgani-
4.	transition		zation
5.	stressor overload	10.	empty nest
6.	resiliency model of	11.	bereavement
	family stress,	12.	bereavement
	adjustment, and		
	adaptation		

True-False

1.	T	8.	F
2.	T	9.	T
3.	F	10.	T
4.	T	11.	F
5.	F	12.	F
6.	T	13.	T
7.	F	14.	T

Multiple Choice

1.	c	14.	d
2.	b	15.	a
3.	b	16.	a
4.	c	17.	d
5.	c	18.	c
6.	a	19.	c
7.	c	20.	b
8.	c	21.	d
9.	d	22.	c
10.	b	23.	a
11.	d	24.	a
12.	c	25.	d
13.	a	26.	b
		27.	a

CHAPTER 15
Divorce

CHAPTER OVERVIEW

This chapter examines the topic of divorce in the
United States. After exploring some of the causes
for the increased divorce rate, the text examines
how married persons weigh divorce against its
alternatives. The six stations of divorce are
explained, along with the economic consequences of
divorce, the relationship between divorce and
children, and the distinction between "his"
divorce and "her" divorce.

Chapter Summary

Divorce rates have risen sharply in this century,
and divorce rates in the United States are now the
highest in the world. The _____ divorce
rate is the number of divorces per 1,000 married
women over age 15. In the past decade, however,
divorce rates have begun to level off.
     Reasons why more people are divorcing than in
the past are related to changes in society:
economic interdependence and legal, moral, and
social constraints are lessening; expectations for
intimacy are increasing; and expectations for
permanence are declining. People's personal
decisions to divorce, or to redivorce, involve
weighing marital complaints--most often problems
with communication or the emotional quality of the
relationship--against the possible consequences of
divorce. Two consequences that receive a great
deal of consideration are how the divorce will
affect children, if there are any, and whether it
will cause serious financial difficulties.
     The divorce experience is almost always far
more painful than people expect. Bohannan has
identified six ways in which divorce affects
people. These six stations of divorce are the
_____ divorce, the legal divorce, the
community divorce, the psychic divorce, the
_____ divorce, and the coparental
divorce. The _____ divorce involves a

healing process that individuals must complete before they can fully enter new intimate relationships.

The _____ divorce is typically more disastrous for women than for men, and this is especially so for custodial mothers. Over the past ten years, child support policies have undergone sweeping changes, which are only now beginning to result in evaluation research.

Husbands' and wives' divorce experiences, like husbands' and wives' marriages, are different. Both the overload that characterizes the wife's divorce and the loneliness that often accompanies the husband's divorce, especially when there are children, can be lessened in the future by more androgynous settlements. Divorce counseling can help make the experience less painful. Joint custody offers the opportunity of greater involvement by both parents, although its impact is still being evaluated. So also is the effect of parents' divorce on children's marital prospects.

Key Terms

You should be able to explain the concepts listed below. In your explanation, try to avoid using the concept you are explaining. You should be able to give several examples of each concept and to explain why each example is an example.

refined divorce rate
  (p. 531)
structured separation
  (p. 540)
readiness for divorce
  (p. 540)
emotional divorce
  (p. 540)
divorce counseling
  (p. 541)
legal divorce
  (p. 541)

entitlement
  (p. 550)
custody
  (p. 558)
child snatching
  (p. 561)
uniform child-custody act
  (p. 561)
joint custody
  (p. 561)
child support
  (p. 564)

Point to Ponder: Some welcomed "dissolution of marriage" statutes, which allowed spouses to petition the court for a dissolution of the marriage, as long as the husband and the wife agreed on disposition of assets, custody, and support. But some research indicates that wives accept dissolution agreements that are not in their long-term best interests. If wives didn't realize at the time that the agreements were not in their best interests, do you think they should later have to live up to those agreements?

Completion

Complete the following sentences by selecting the correct alternative from the terms above. Some may be used more than once. Some may not be used at all. Filling in a blank may require more than one word.

1. In _____, partners try with the help of marriage counselors to negotiate various conflicts, grievances, and misunderstandings.

2. Court-ordered _____ does have more negative outcomes than other forms of this divorce related issue.

3. In _____, intensely hostile couples live apart for a limited time, avoid resorting to lawyers, and continue involvement in counseling, after which time they ordinarily know whether or not they want to continue to seek a divorce.

4. In _____, a third party--often a lawyer-therapist team--confers with the couple to produce an arrangement that is suitable to the divorcing couple and that has their agreement.

5. Former in-laws, for example, illustrate _____.

6. When the state dissolves the marriage by means of a court order declaring the marriage at an end, it is appropriate to speak of a(n) _____.

7. When spouses no longer engage in bonding or communication but instead engage in alienating feelings and behavior, then it is appropriate to speak of the occurrence of _____ _____.

8. The refined divorce rate is the number of divorces per _____.

9. _____ refers to the assumption of primary responsibility for making decisions about the children's upbringing and general welfare and for the carrying out of those decisions.

10. _____ is a procedure that does not seek to assign blame but instead addresses an appropriate resolution to the marriage situation.

11. In most states _____ had been considered a misdemeanor and is called "custodial interference."

12. The text defines a _____
    as a woman who has been a full-time housewife
    and who may have been providing support for
    her and, possibly, for their children.

13. In _____
    situations, both divorced parents continue to
    take equal responsibility for important
    decisions about the child's general
    upbringing.

14. In a(n) _____,
    the finances of the partners are separate,
    distinct entities or units.

15. In _____, the
    divorced father usually takes legal
    responsibility for financial support and the
    divorced mother ordinarily makes decisions
    about the children's upbringing and general
    welfare.

16. When the divorced person has "distanced"
    herself/himself from the previous spouse, has
    gained autonomy, and feels like a whole
    person again, it is appropriate to use the
    concept of _____ to
    refer to the situation.

17. The term _____ does
    not refer to something that is given to a
    person but instead refers to something that
    the person is given because it has already
    been earned. Betty Friedan equates it to
    "severance pay" for work done at home during
    the marriage.

18. When the ex-husband provides the money for
    the ex-wife to attend school, to be trained
    or retrained, and to find employment, then it
    is appropriate to use the term
    _____.

Key Research Studies

You should be familiar with the main question being investigated and the research results for the following studies:

Weitzman: The Divorce Revolution
Wallerstein: stresses for the children of divorce
Amato and Keith: children's postdivorce adjustment
Fisher: noncustodial mothers

Key Theories

Paul Bohannon: six stations of divorce

True-False

____ 1. The number of divorces per year takes into account the general increase in population.

____ 2. The ratio of current marriages to current divorces is a faulty measure.

____ 3. The crude divorce rate is the number of divorces per 1,000 population.

____ 4. The gap between blacks and whites has increased the most for the less well educated, and this has implications for divorce rates.

____ 5. The higher the income, the less likely couples are to divorce.

____ 6. Intriguingly enough, recent evidence indicates that divorce can be an uplifting, fulfilling experience--bordering on the romantic.

_____ 7. Divorced and separated individuals have lower levels of life satisfaction and a more negative general mood than married people.

_____ 8. Children whose parents divorce will experience a noticeable drop in the amount of income available to provide for them.

_____ 9. At ten years after divorce, Wallerstein's "Children of Divorce" did not continue to experience a feeling of loss insofar as having an intact family was concerned

_____ 10. Wallerstein found that girls with divorced parents are better adjusted than are boys--at least in the long run.

_____ 11. Redivorces take place more rapidly than first divorces.

_____ 12. Recently, marriage counselors have begun to repudiate divorce counseling, urging partners instead to seek the advice of accountants and human service workers.

_____ 13. Annulment legally states that a marriage once existed but no longer does.

_____ 14. In legal separation, the couple remains married, but separate residences are recognized by the court.

_____ 15. All of the states in the United States now have no-fault divorce, or dissolution of marriage.

_____ 16. The divorced wives interviewed by Wallerstein usually plunged into educational and professional self-development.

_____ 17. There are six stations to Bohannan's analysis of divorce.

_____ 18. The "psychic" divorce is one of the various stations of divorced discussed by Bohannan.

_____ 19. Though there are no divorce counselors at present, the text argues that this is an occupational category that is needed and should exist.

_____ 20. No-fault divorce is a type of divorce in which the participating spouses are essentially combatants in a "legal arena."

_____ 21. Rehabilitative alimony is a relatively new category of alimony granted to previous spouses--male or female--who face medical expenses incurred as a direct or indirect result of the marital relationship.

Multiple Choice

1. In the twentieth century, the frequency of divorce exhibits dips and upswings surrounding:
   a. major wars
   b. whether or not divorce is an "in" thing at the moment
   c. passage of major legislation making changes in the costs/benefits of divorce
   d. the Roman Catholic and Jewish attitudes toward annulment, dissolution, and divorce

2. About ____ of all households are headed by men, and _____.
   a. 3 percent; this does not vary by race
   b. 11 percent; most are white
   c. 15 percent; most are black
   d. 23 percent; most are black

3. According to the text, the higher the _____, the lower the _____.
   a. altitude; fertility
   b. fertility rate; divorce rate
   c. social integration; moral integration
   d. social class; divorce rate

4. According to the text, about _____ of all black children born to a marriage will experience the dissolution of that marriage.
   a. one-sixth
   b. one-third
   c. two-thirds
   d. five-sevenths

5. Class differences in divorce rates:
   a. are primarily religious differences
   b. reflect personality differences by class
   c. reflect philosophical differences about marriage
   d. have been narrowing in recent years

6. Which of these is NOT one of the fifteen suggestions for healing after a divorce?
   a. Be gentle on yourself.
   b. Remember that it's okay to feel depressed.
   c. Try to find an all-consuming passionate romance or a new project that requires great time and energy.
   d. If possible, don't take on new responsibilities.

7. According to the text, children of divorce feel less protected:
   a. economically
   b. politically
   c. intellectually
   d. psychologically

8. Which of the following is one of the stations of divorce as recognized by Paul Bohannan?
   a. the mutual divorce
   b. the children's divorce
   c. the coparental divorce
   d. the culminating divorce

9. Beginning to date early after divorce:
   a. has more negative than positive consequences
   b. is usually the norm in the western and southwestern states
   c. is associated with higher rates of psychosomatic illness
   d. may disrupt relationships with one's children

10. In most states, _____ has been defined as a misdemeanor.
    a. obtaining a divorce
    b. refusal to acknowledge divorce proceedings
    c. child snatching
    d. nonpayment of alimony

11. To be successful a(n) _____ requires a long period of mourning.
    a. mortality divorce
    b. divorce that has been a long time in coming
    c. economic divorce
    d. psychic divorce

12. The increase in single families has multiple causes, but the major cause is:
    a. divorce
    b. rising unwed birthrates
    c. greater tendency for unmarried mothers to establish independent households
    d. decline in the number of 2- and 3-bedroom apartments

13. The "refined divorce rate" is:
    a. the rate of divorces for persons who leave the entire matter for their attorneys; the persons themselves do not show up in court
    b. the divorce rate that has been combed and recombed to eliminate from the database divorces that were actually separations and dissolutions
    c. the number of divorces per 1,000 married women over age 15
    d. the number of divorces per 1,000,000 marriages, regardless of marriage type-- civil marriage, religious marriage, or common law

14. A _____ study of divorce would have to study a cohort of married couples over a lifetime, collecting and analyzing data for the entire period.
    a. longitudinal
    b. randomized
    c. stratified
    d. time-intensive

15. The prediction for the near-future is that
    _____ of first marriages taking place today
    will end in divorce.
    a. one-quarter
    b. one-half
    c. two-thirds
    d. three-fourths

16. The high divorce rate:
    a. does not mean Americans have given up on
        marriages; they just want rewarding ones
    b. now stands poised at just under one-half
        of all marriages
    c. is probably an overestimate since it
        includes dissolutions of marriage in
        divorce statistics
    d. reflects the fact that most divorces are
        in the middle years of marriage

17. _____ take place more rapidly than _____.
    a. first remarriages; second or third
        remarriages
    b. redivorces; first divorces
    c. African-Americans' divorces; other
        Americans' divorces
    d. upper-class divorces; working-class
        divorces

18. Statistics show that divorce rates have risen
    as:
    a. women's employment opportunities have
        increased
    b. the migration back to city fringes
        produced greater acceptability of divorce
    c. the economic crunch has eroded husbands'
        ability to support families--probably the
        greatest cause of the increased rates
    d. the Catholic Church has dropped its ban
        against the procedure

19. Most divorces occur:
    a. relatively early in marriage: the first
        five years
    b. when the partners' children are teenagers
    c. after the last child leaves home
    d. during the partners' middle age

20. No-fault divorce legally abolishes:
    a. the need for expensive legal proceedings
    b. legal proceedings at all; the couple simply go their separate ways
    c. the concept of the "guilty party"
    d. marriage ties, but does not abolish economic ties

21. Psychic divorce:
    a. requires dismissing the other partner from one's mind, right from the start
    b. requires a period of mourning
    c. has causes that lie primarily in the area of mental health
    d. is divorce that was foreseen by one or both of the partners, usually about five to ten years before divorce was actually spoken about openly

22. Displaced homemakers are divorced women who:
    a. literally find themselves out on the street, looking for new residence
    b. find that someone else has taken their place in their ex-husband's affections
    c. are older, full-time homemakers who suddenly find themselves divorced and without adequate support
    d. are not given custody of their children; that custody goes to the husband and/or to the husband's new partner

23. The "sleeper effect" is found among:
    a. girls coming from divorced families
    b. divorced wives
    c. teenage boys coming from divorced families
    d. the time of expenses related to getting divorced

24. About ___ of couples who divorce have children under age 18.
    a. 10 percent
    b. 35 percent
    c. 53 percent
    d. 68 percent

25. Adult children of divorce express:
    a. more accepting attitudes toward divorce
    b. deep resentment and hostility toward the noncustodial parent
    c. resentment toward one biological parent, but not usually toward both
    d. a desire to remain single

Your Opinion, Please: Imagine two people whose relationship culminated in marriage. They loved each other, felt the need for a shared future, and seemed perfectly suited for each other--and married. But people change. And not even marriage stops change. Often people who were suited to each other change in the same direction and remain well-matched. But sometimes, people change in different directions so that one or both no longer want the same things and no longer feel comfortable with the idea of a shared future. If one partner changes in a direction that is difficult or unacceptable to the other partner, should they face the fact and seek a divorce? Or should they work at the relationship? Should one partner suppress their new direction for the sake of the current partnership or should they divorce rather than compromise in any way? Can you think of some concrete examples?

Short-Answer Essay Questions

The following are sample short-answer essay questions--questions of the type you may be asked if your instructor uses questions like these. Even if your instructor does not use questions like these, you can help organize and consolidate your learning if you can answer these questions in a <u>well-organized</u> and <u>complete</u> manner. Do not think that brief essays are easier than lengthy essays. They may be more challenging because you are required to be brief yet complete. And after you have answered these, try making up some

similar questions and then answering them.
Suggestion: The Study Questions at the end of the
text chapter make very good short-answer
questions.

1. List five general reasons for the increased
    divorce rate. For each, write an example,
    making sure it is clear <u>why</u> the example is a
    good example.

2. Distinguish between what Bohannan calls "the
    legal divorce" and "the community divorce."

3. Explain the concept of "rehabilitative
    alimony."

Essay

The following are sample essay questions--
questions of the type you may be asked if your
instructor uses essay questions. Even if your
instructor does not use essay questions, you can

help organize and consolidate your learning if you can answer these questions in a <u>well-organized</u> and <u>complete</u> manner. Usually, the third essay question is the most challenging.

1. More couples are divorcing or in other ways terminating their marriage. Explain why this is so.

2. Material in the text suggests that many divorces go through various stages. Though it may seem unique to the persons involved, sociologists notice some of the same stages when studying many divorces. What are these stages? Briefly describe each stage in the process.

3. Louise and Andy considered divorcing for some time, and eventually got divorced. Write an essay in which you explore how people come to such a decision and what usually happens to them as they divorce. Explore briefly any ways in which Louise's divorce might be different from Andy's divorce.

<hr>

Answers

Chapter Summary

the **<u>refined</u>** divorce rate
the **<u>emotional</u>** divorce
the **<u>economic</u>** divorce, and the coparental divorce
the **<u>psychic</u>** divorce involves
the **<u>economic</u>** divorce is typically more disastrous

Completion

1. divorce counseling	11. child snatching
2. custody	12. displaced homemaker
3. structured separation	
4. divorce mediation	13. joint custody
5. relatives of divorce	14. economic divorce
6. legal divorce	15. co-parental divorce
7. emotional divorce	
8. 1,000 married women over age 15	16. psychic divorce
	17. entitlement
9. custody	18. rehabilitative alimony
10. no-fault divorce	

True-False

1.	F	11.	T
2.	T	12.	F
3.	T	13.	F
4.	T	14.	T
5.	T	15.	T
6.	F	16.	F
7.	T	17.	T
8.	T	18.	T
9.	F	19.	F
10.	F	20.	F
		21.	F

Multiple Choice

1.	a	13.	c
2.	a	14.	a
3.	d	15.	c
4.	c	16.	a
5.	d	17.	b
6.	c	18.	a
7.	a	19.	a
8.	c	20.	c
9.	d	21.	b
10.	c	22.	c
11.	d	23.	a
12.	a	24.	c
		25.	a

CHAPTER 16
Remarriage

Chapter Overview

This chapter examines remarriage in the United
States. The concept of remarriage is examined
from a theoretical perspective, and some basic
demographic facts about remarriage are summarized.
The text reviews research findings about remarried
spouses' satisfaction with remarriage and pays
special attention to the fact that remarriage is
in many ways a "normless norm." After reviewing
what is known about stepparenting, the chapter
ends with some observations about writing a
personal remarriage contract.

Chapter Summary

Although remarriages have always been fairly
common in the United States, patterns have
changed. Remarriages are far more frequent now
than they were earlier in this century, and they
more often end in divorce than in widowhood. The
courtship process by which people choose partners
has similarities to courtship preceding first
marriages, but the basic exchange often weighs
more heavily against older women, and homogamy
tends to be less important.

Second marriages are usually about as happy
as first marriages, but they tend to be slightly
less stable. An important reason is the lack of a
cultural script. Relationships in immediate
remarried families and with kin are often complex,
yet there are virtually no social prescriptions
and legal definitions to clarify roles and
relationships.

The lack of cultural guidelines is clearest
in the stepparent role. Stepparents are often
troubled by financial strains, role ambiguity, and
stepchildren's hostility. The contradictions in
expectations for the role of stepmother are
sufficiently great that Phyllis Raphael refers to
this configuration of conflicts as the stepmother
_____. For male stepparents, one

of the first difficulties a stepfather encounters is the _____.   Marital happiness and stability in remarried families are greater when the couple have strong social support, high expressiveness, a positive attitude about the remarriage, low role ambiguity, and little belief in negative stereotypes and myths about remarriages or stepfamilies. Personal remarriage agreements can help to establish an understanding where few social norms exist.

_____

Key Terms

You should be able to explain the concepts listed below. In your explanation, try to avoid using the concept you are explaining. You should be able to give several examples of each concept and to explain _why_ each example is an example.

double remarriages
   (p. 581)
single remarriages
   (p. 581)
remarried families
   (p. 583)
quasi-kin
   (p. 585)

coparenting
   (p. 586)
stepmother trap
   (p. 591)
hidden agenda
   (p. 594)

Point to Ponder: According to the text, remarriages have a higher divorce rate than first marriages. The text lists several possible reasons for the higher divorce rate for remarriages:

1.   the social class of people who divorce and remarry
2.   persons who remarry may be accepting of divorce--they have done it before
3.   remarriages have stresses in addition to the usual stresses of marriage, including the presence of stepchildren

Can you think of **other** reasons? Please give it some thought, and write out the reasons that occur to you.

Completion

Complete the following sentences by selecting the correct alternative from the terms listed above. Some may be used more than once. Some may not be used at all. Filling a blank may require more than one word.

1. Conflicting expectations for behavior and attitude are the essential ingredients for the _____.

2. The _____ consists of partners in a remarriage failing to fully reveal their expectations for the partner in the remarried relationship.

3. Bohannan suggests the term _____ to refers to the person one's former spouse remarries. The prefix of the word actually means "somewhat" or "similar to" when used along with a suffix.

4. In _____ both partners have been married before.

5. Members of _____ have no cultural script.

6. In a _____, one partner is marrying for the first time and the other partner is remarrying.

7. In a _____, both partners are remarrying.

Key Research Studies

You should be familiar with the main question being investigated and the research results for the following studies:

    Rodgers and Conrad: courtship for
        remarriage
    Spanier and Furstenberg: well-being among
        the remarried

Bohannan; Cherlin: general problems of
   stepparenting
Raphael: strains of the stepmother trap
U.S. National Center for Health Statistics:
   various data about remarried individuals
   and couples

Key Concepts

Bohannan: quasi-kin
Bohannan: role ambiguity in stepparenthood
Raphael: the stepmother trap

True-False

____ 1. Remarriage is an alternative that more
          Americans are choosing.

____ 2. The U.S. Supreme Court has ruled that
          failure to pay child support is not by
          itself enough to prevent a remarriage.

____ 3. The average divorced person who remarries
          does so within six months to one year
          after the divorce.

____ 4. Remarriages differ from first marriages
          in important ways.

____ 5. In general, marriage favors husbands
          more than wives.

____ 6. The "double standard of aging" works **for**
          women rather than **against** them in the
          remarriage market.

____ 7. A factor that works against women in the
          remarriage market is the presence of
          children.

____ 8. The divorce rate has leveled off, but
          the remarriage rate has declined
          steadily since 1966.

_____ 9. Generally, people who remarry after divorce are less willing to choose divorce as a way of resolving an unsatisfactory marriage.

_____ 10. Researchers have found that the presence of stepchildren is not a significant factor in the instability of marriages.

_____ 11. Quasi-kin are kin whose deviant behavior has resulted in distancing them from other family members.

_____ 12. The role of stepparent is less clearly defined than the role of parent in our society.

_____ 13. Stepchildren tend to be well adjusted but do not get along with their stepfather as well as other children do with their natural fathers.

_____ 14. The so-called "hidden agenda" is a contradiction in terms, since it occurs when the participants have been **too** clear and **too** candid about their expectations for the remarriage relationship.

_____ 15. People who remarry would be wise to check things out with a lawyer.

Multiple Choice

1. According to the text, the remarriage rate peaked most recently during which of the following periods?
   a. at the end of World War II
   b. at the end of the Great Depression
   c. in 1989
   d. in 1966

2. According to the text, ex-husbands more than ex-wives may want to:
   a. request adoption of their stepchildren
   b. avoid marriage if the stepchildren are already adults
   c. remarry
   d. qualify for "palimony"

3. There is some evidence that _____ face more opposition to remarriage.
   a. upper middle-class people
   b. older people
   c. people who are in love
   d. people who are authoritarian

4. Remarriages (marriages in which at least one partner had previously been divorced or widowed) now comprise ____ percent of all marriages.
   a. 13
   b. 26
   c. 34
   d. 46

5. Remarriages tend to be _____ as compared with first marriages.
   a. more practical
   b. less practical
   c. more truly romantic
   d. less stable

6. According to the text, our culture has no fixed norms about how to deal with:
   a. spouses' previously incurred financial obligations
   b. extreme endogamy
   c. ex-spouses
   d. quasi-kin

7. Today, the largest proportion of the remarried population are divorced people who have married:
   a. widows or widowers
   b. someone in a social class lower than their own
   c. persons who have never been married before
   d. other divorced people

8. Which of these is analyzed as a "trap" in the chapter on remarriage?
   a. stepgrandmotherhood
   b. stepgrandfatherhood
   c. stepfatherhood
   d. stepmotherhood

9. According to the text, _____ are important because society has not yet evolved an effective cultural model for these complex relationships.
   a. filial responsibility laws
   b. paternal responsibility laws
   c. remarriage contracts
   d. compacts between nonrelated stepchildren

10. One consequence of _____ is that society expects acquired parents and children to love one another in much the same way as biologically related parents and children do.
    a. modern familistic involvement
    b. role ambiguity
    c. status complexity
    d. legal requirements

11. A reason for writing a remarriage contract is that second-time spouses _____ even more than do spouses in first marriages.
    a. may feel the conflict taboo
    b. may doubt their love
    c. may have more faith in their love
    d. are likelier to marry outside civil law

12. Jessie Bernard pointed out that as a whole, married life tends to:
    a. reduce the number of life-spheres controlled by parents and increase the number of life-spheres influenced by offspring
    b. intensify or focus the praise/blame ratio on either spouse
    c. place greater stresses on women who become mothers and reduce stresses on husbands
    d. reduce psychological stressors for women and increase economic and psychological stressors for both sexes

13. Which of the following is TRUE? Regarding the remarriage rate, ...
    a. the African-American rate is nearly twice that of either the Asian-American rate or the Mexican-American rate
    b. women's remarriage rate is less than half that of men's
    c. it is highest during the winter months and lowest during the autumn months
    d. the remarriage rate has been declining steadily since 1955

14. In second or later marriages, does homogamy affect choice of marriage partner or level of happiness of the marriage? Evidence indicates that:
    a. it does not
    b. there is no particular pattern that is statistically supportable
    c. it does support the homogamy hypothesis
    d. psychological need overrides experiential learning

15. According to the text, which marriages tend to be less homogamous?
    a. younger marriages
    b. older marriages
    c. marriages of the upwardly mobile
    d. marriages of the well educated

16. Regarding first and later marriages, researchers consistently find little difference in:
    a. number of offspring
    b. level of emotional bonding between spouses
    c. level of marital happiness
    d. probability of future divorce

17. According to research dating back to the 1950s, remarriages are ____ more likely to end in divorce than are first marriages.
    a. no
    b. 10 percent
    c. 35 percent
    d. 55 percent

18. On the whole, remarriages are:
    a. more productive of high-achieving offspring
    b. less productive of incidents of conflict
    c. more unstable than first marriages
    d. likelier to produce higher levels of marriage satisfaction, mainly due to the amount of time the partners can spend together

19. Given _____, many older women may come to prefer a single life-style.
    a. the sex ratio during the older years
    b. decreasing tolerance of others' eccentricities
    c. the wide variety of economic incentives
    d. the deeply felt social pressures not to remarry

20. "Uncle" Fred (who is not an uncle biologically), "Aunt" Denise (who is not an aunt biologically), who have this relationship through remarriage of a spouse, are examples of:
    a. quasi-kin
    b. illegitimate kin
    c. out-of-bounds kin
    d. referent kin

Your Opinion, Please: Some people remarry and re-divorce several times. The divorce rate is as high as it is partly because it reflects multiple remarriages and divorces. In an earlier "Point to Ponder," you were asked for reasons why people who divorce are likelier to divorce again. Why, in your opinion, does this happen? Be specific about the reasons you give. Do you think these reasons are the same reasons as for first marriages and divorces, or do you think the reasons are different?

Short-Answer Essay Questions

The following are sample short-answer essay questions--questions of the type you may be asked if your instructor uses questions like these. Even if your instructor does not use questions like these, you can help organize and consolidate your learning if you can answer these questions in a __well-organized__ and __complete__ manner. Do not think that brief essays are easier than lengthy essays. They may be more challenging because you are required to be brief yet complete. And after you have answered these, try making up some similar questions and then answering them. Suggestion: The Study Questions at the end of each text chapter make very good short-answer essay questions.

1. What have been remarriage trends since 1900?

2. What are the differences between remarriage in which only one person has been married before and remarriages in which both persons have been married before?

3. Ex-husbands more than ex-wives want to remarry. Why?

Essay

The following are sample essay questions--
questions of the type you may asked if your
instructor uses essay questions. Even if your
instructor does not use essay questions, you can
help organize and consolidate your learning if you
can answer these questions in a well-organized and
complete manner. Usually, the third essay
question is the most challenging.

1. Explore the extent to which remarriages are
    happy and stable.

2. Remarried couples frequently state that it
    seems difficult for them to "fit in" and that
    they often don't know what's expected of them.
    According to the text, why is this so?

3. Imagine that a person you know is getting
    remarried and wants to know what to expect.
    At lunch with this friend, you explain the
    things you think are most important to know
    about remarriages. What do you tell your
    friend? Give a well-organized answer, and be
    complete.

Answers

Chapter Summary

the **stepmother** trap
is the **hidden** agenda

Completion

1. stepmother trap
2. hidden agenda
3. quasi-kin
4. double remarriage
5. remarried families
6. single remarriage
7. double remarriage

True-False

1.	T	9.	F
2.	T	10.	F
3.	F	11.	F
4.	T	12.	T
5.	T	13.	F
6.	F	14.	F
7.	T	15.	T
8.	T		

Multiple Choice

1.	d	11.	a
2.	c	12.	c
3.	b	13.	b
4.	d	14.	a
5.	d	15.	b
6.	d	16.	c
7.	d	17.	b
8.	d	18.	c
9.	c	19.	a
10.	b	20.	a